Real-Time
Medical Image
Processing

Real-Time Medical Image Processing

EDITED BY

Morio Onoe

University of Tokyo
Tokyo, Japan

Kendall Preston, Jr.

Carnegie-Mellon University
Pittsburgh, Pennsylvania

AND Azriel Rosenfeld

University of Maryland
College Park, Maryland

PLENUM PRESS · NEW YORK AND LONDON

Library of Congress Cataloging in Publication Data

Main entry under title:

Real-time medical image processing.

   Based on proceedings of the Japan-United States Seminar on Research Towards Real-Time Parallel Image Analysis and Recognition, Oct. 31—Nov. 4, 1978, Tokyo, Japan.
   Includes bibliographies and index.
   1. Imaging systems in medicine—Congresses. 2. Image processing— Congresses. 3. Real-time data processing—Congresses. I. Onoe, Morio. II. Preston, Kendall, 1927-   III. Rosenfeld, Azriel, 1931-   IV. Japan-United States Seminar on Research Towards Real-Time Parallel Image Analysis and Recognition, Tokyo, 1978. [DNLM: 1. Diagnosis, Computer assited—Congresses. WB141 J35r 1978]
R857.06R42   610'.28'54   80-23779

Based on proceedings of part of the Japan — United States Seminar on Research Towards Real-Time Parallel Image Analysis and Recognition, held in Tokyo, Japan, October 31, 1978—November 4, 1978.

© 1980 Plenum Press, New York
Softcover reprint of the hardcover 1st edition 1980
A Division of Plenum Publishing Corporation
227 West 17th Street, New York, N.Y. 10011

ISBN 978-1-4757-0123-4   ISBN 978-1-4757-0121-0 (eBook)
DOI 10.1007/978-1-4757-0121-0

PREFACE

In order to realize real-time medical imaging systems, such as are used for computed tomography, automated miscroscopy, dynamic radioisotope imaging, etc., special technology is required. The high-speed image source must be successfully married with the ultra-high-speed computer. Usually the ordinary general-purpose computer is found to be inadequate to the image generation and/or image processing task. The ordinary computer executes instructions at between 1 and 10 million per second. Speed has improved by only about a factor of 10 during the past 20 years. In contrast a typical computer used in recognizing blood cell images at 10,000 per hour must execute instructions at between 1 billion and 10 billion per second. Similar execution rates are required to construct a computed tomography image in real-time (1 to 10 seconds).

For the reasons given above, engineering development in image generation and processing in the field of biomedicine has become a discipline unto itself; a discipline wherein the computer engineer is driven to design extremely high-speed machines that far surpass the ordinary computer and the x-ray, radioisotope, or microscope scanner designer must also produce equipment whose specifications extend far beyond the state-of-the-art.

Recognizing the importance of this new field the editors of this book organized the Japan-United States Seminar on Research Towards Real-Time, Parallel Image Analysis and Recognition which was held from October 31st to November 4th in Tokyo in 1978. The chapters presented in this book represent the major contributions of the participants in this seminar updated to reflect changes in the state-of-the-art during the interval 1978-1980. The book is divided into three sections. Section 1 presents two chapters concerned with a broad spectrum of medical image analysis. Sections 2 and 3 concentrate on applications in the analysis of radiological images and of cell images, respectively. The names and addresses of all authors are furnished in an appendix.

     The editors wish to acknowledge the support of the Society for the Promotion of Science of Japan and of the United States National Science Foundation under the auspices of the Japan-United States Cooperative Science Program. Also acknowledged are the typing services of Caroline Wadhams' organization (Report Production Associates, Cambridge, Massachusetts) and her staff, particularly Susan Dunham, who produced the manuscript. Also acknowledged are the efforts of Judy Tyree's organization (Executive Suite, Tucson, Arizona) and Rosalie Ballentine of her staff who assisted with the subject index. The cooperation of the staff of Plenum Press, in particular James Busis (Editor) and John Matzka (Managing Editor) is greatly appreciated.

M. Onoe
K. Preston, Jr.
A. Rosenfeld

CONTENTS

GENERAL

Towards an Image Analysis Center for Medicine
T. Kaminuma, J. Kariya, I. Suzuki and S. Kurashina

Cellular Computers and Biomedical Image Processing
S. R. Sternberg

RADIOLOGY

HISTOLOGY AND CYTOLOGY

REAL-TIME MEDICAL IMAGE PROCESSING

TOWARDS AN IMAGE ANALYSIS CENTER FOR MEDICINE

T. Kaminuma, J. Kariya, I. Suzuki, and S. Kurashina

Tokyo Metropolitan Institute of Medical Science

Tokyo, JAPAN

1.   INTRODUCTION

Regardless of the success of computerized tomography and white blood cell differentiation as discussed by Onoe (1976) and Preston (1976) the effectiveness of the application of computers to other tasks in medical image processing is still questionable (see Agrawala, 1977). Great research efforts are still needed if one wants to bring significant advances in daily medical services through the application of image processing technology.

It is costly and inefficient for each medical researcher to construct his own image processing facility. The TIMS (Tokyo Metropolitan Institute of Medical Science) image analysis systems are designed as powerful central tools for multi-user requests for research in medical image analysis by a multi-disciplinary research unit dedicated to computer applications in medicine. The project is being carried out jointly between the epidemiology and medical engineering divisions as described in Kaminuma (1978). One of the major current research activities of the group is to construct comprehensive medical data analysis systems including statistical analysis, time series analysis, and image and picture analysis.

The analysis systems aim to provide data bases for accumulating relevant medical data and subroutine libraries by which medical personnel, who are not computer specialists, can analyze various features of their data with ease. It is expected that the TIMS image analysis system will play the same role as today's "computing center" for medical image processing tomorrow. It is intended that the facility will allow medical personnel to process any medical

1

picture data, without knowing details of the hardware or even the special software of the system.

The system consists of an interactive image analysis system and a computerized microscope. The interactive image analysis system plays the central role in picture processing. It is expected to be operated rather independently from the measuring apparatus through which image data are created. Inputs to this system are limited either to films or to digitized pictures stored in storage media. On the other hand the computerized microscope is used to scan, preprocess, and store microscope images.

2.   THE INTERACTIVE IMAGE ANALYSIS SYSTEM

The system has two input units, a TV camera and drum scanner, an interactive image display, and several standard computer peripheral devices. Figure 1 shows the block diagram of the system.

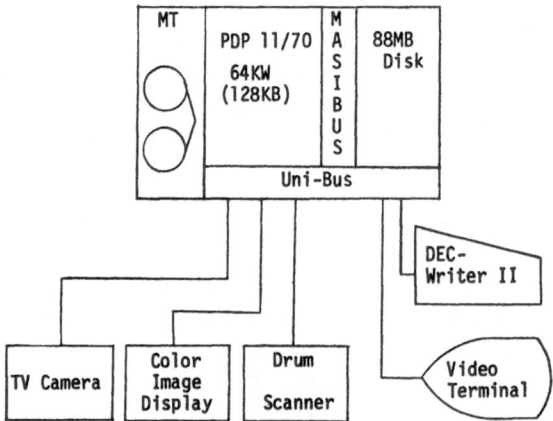

Fig. 1. Block diagram of the interactive image analysis system.

2.1 The Input Units

Images are digitized either by the TV camera system or by the drum scanner densitometer. The TV camera system can easily handle object films. It permits immediate investigation of object features. Its accuracy, dynamic range, and spatial resolution are not as good as those of the drum scanner. The disadvantage of the latter is its slow scanning speed. These two input units are complementary in their functions. We consider the drum scanner for experimental purposes and the TV camera for routine operations.

2.1.1 *The TV Camera Input System*

This system is a NAC Co. model VD-520A. It consists of a
light box, the TV camera, a shading correction circuit, and A/D
converter (Figure 2). Each picture frame is resolved into 512 x
480 pixels at 8 bits per pixel. The shading correcter corrects
overall shading inhomogeneity due to the camera, optical lenses,
and the illumination of the light box. Temporal fluctuations are
smoothed out by averaging for 10 to 100 times the repeated measurement values. The input image is monitored by a TV monitor. The
grayness value of a pixel is selected by the crossing point of two
lines in X and Y. The grayness distribution along the Y line can
also be plotted on the monitor. By rotating three filters in front
of the camera, three different color components will be merged
into one color picture by the computer.

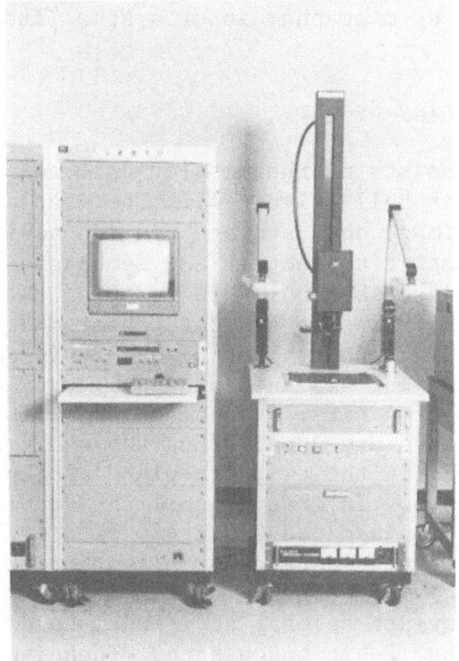

Fig. 2. The television camera input system (NAC model VD-520A).

2.1.2 *The Drum Scanner*

The drum scanner is manufactured by the Abe Co. (Model 2604).
See Figure 3. Its specification is almost the same as that of the
Optronix P-1000 or Joyce-Loeble scanner model 50. The optical

density range is 0-4D, aperture 25μ-1000μ, and sampling pitch 25μ, 50μ, and so on. When using the color option, the three color components with the black and white component are changed into four one-byte samples. These data can be sent to the CPU either in byte format or as 2 words (4 bytes) simultaneously.

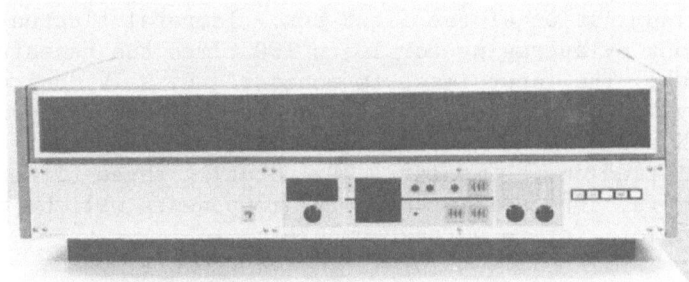

Fig. 3.   The drum scanner input system (Abe model 2604).

2.2   The Image Display System

The system consists of the display unit and a keyboard with a joystick and a track ball (Figure 4). The display unit has 16 memory planes for image data storage and two additional memory planes for alphanumeric characters and symbols. The image plane has 320 x 256 addresses but only 320 x 240 of them are actually displayed on the CRT. In the standard color mode the 16 planes are divided into three, four-plane blocks assigned to the three color components leaving 4 planes unused, i.e., each color component consists of 4 bits. For black and white (B/W) images only six memory planes are used. Thus two 6-bit B/W picture frames can be stored in the two, six-plane blocks. Storage modes are shown in Figure 5. In what we call the double accuracy display mode one frame is displayed at 640 x 480 with eight possible color components, i.e., three binary color pictures are merged.

The system provides several additional hardware functions which manipulate the 12-bit image data. These include 12-bit to 12-bit mapping for color modulation. The modulation function may be used for emphasizing a certain color range or changing the color tone. This binary transformation function is defined by software. Picture data are sent from buffer memories to the plane memories in DMA mode in 1 word (16-bit) slices. Within each 16-bit unit, we may rotate the data by any even amount and may mask the bits to save only what we want shown on the screen. See Figure 6. It is also possible to eliminate the pixel whose color components are matched to the levels specified by the dials on the display panel.

Fig. 4. The interactive color image display.

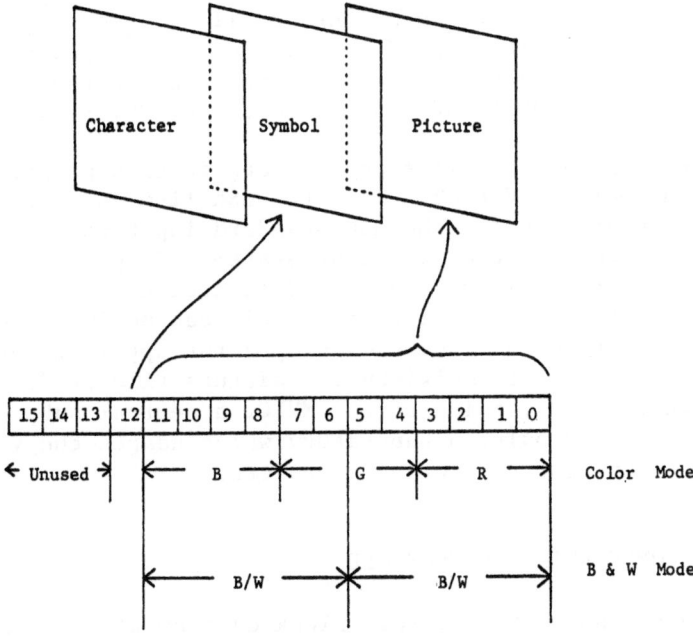

Fig. 5. Diagram of memory plane storage format.

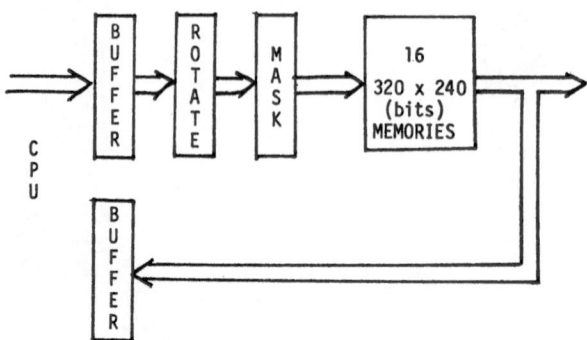

Fig. 6. Flow chart for CPU to memory plane communication.

2.3 The Computer System

The computer is the PDP 11/70 with 64KW main memory, a 88MB magnetic disk, a 800/1600 BPI magnetic tape and two typewriters. One of the typewriters, the DEC Writer II, is used as the console, while the videoterminal VT-52 is used for image operations. The previously explained subsystems are connected to the unibus of the PDP 11/70 through DEC's standard interfaces, DR11-B and DR11-C.

According to our philosophy all application programs must be written in FORTRAN IV PLUS. Thus the RSX 11-M operating system is used and the drivers for the non-standard input and image display devices are added to the operating system. Programmers can control the non-standard devices through QIO (queueing I/O) subroutines or the subroutines which consist of the QIO subroutines. However various subroutines have been developed for applications. A particular example is the multi-buffer picture file handling subprogram which is a version of the MVIO, a central subprogram developed at Jet Propulsion Laboratory which adapts the VICAR system for IBM 360 and the PDP series computers.

3. THE COMPUTERIZED MICROSCOPE

Figure 7 shows the overall block diagram of the computerized microscope.

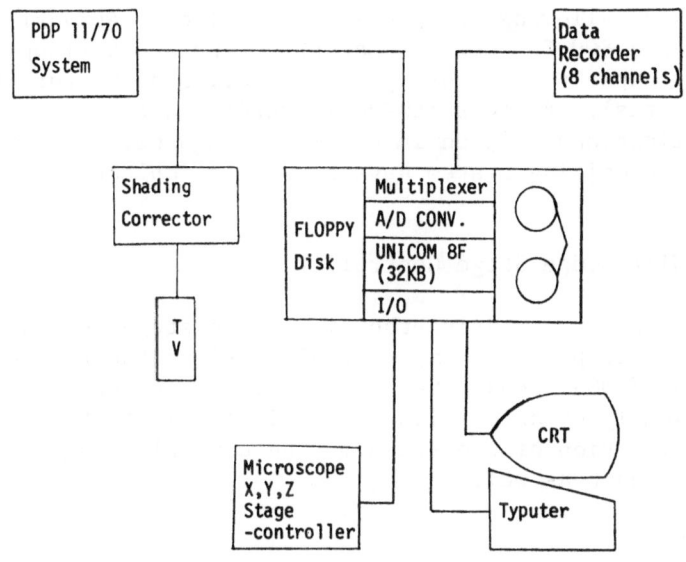

Fig. 7. Block diagram of the computerized microscope.

3.1 The Input System

All image data are transferred by a high resolution TV camera into a digital converter, which consists of a sample and hold circuit, a multiplexer, and an A/D converter with 256-level resolution. Each picture frame is digitized into a 480x480 byte record. These records may be stored on a floppy disk or a magnetic tape. The TV camera is supported by a moveable arm on an oscillation-absorbing table, adjacent to the microscope. (The table also contains the light box for film scanning.) To record color images filters are inserted into the microscope and the TV camera. The TV image can also be sent to the monitor system of the interactive system. Thus it is possible to analyze the microscopic images directly with the PDP system.

3.2 The Computer System

The UNICOM-8F microcomputer manufactured by Unicom Automation, with the Intel 8080 microprocessor as the CPU and 32K byte memory, is used. The computer has an ink jet typewriter with paper tape puncher and reader and a 16x40 character display with numerical

function keys allowing interactive operations. As auxiliary memory
it also has a 800 BPI density magnetic tape and a floppy disk with
the memory capacity to store just one full size picture frame
(480x480 bytes). Basic software libraries have been developed for
image applications written in assembly language. A FORTRAN compiler and an editing system have also been implemented.

3.3 The Microscope Stage Controller

    The computer can drive the microscope stage in X and Y direc-
tion in 0.5μ steps (minimum). It also drives the Z direction
(focus) in 1/800μ steps. See Figure 8. The stage can also be
driven manually by an off-line control unit in fast and slow modes.
The X and Y motion of the stage may be controlled by joysticks on
the control unit as well.

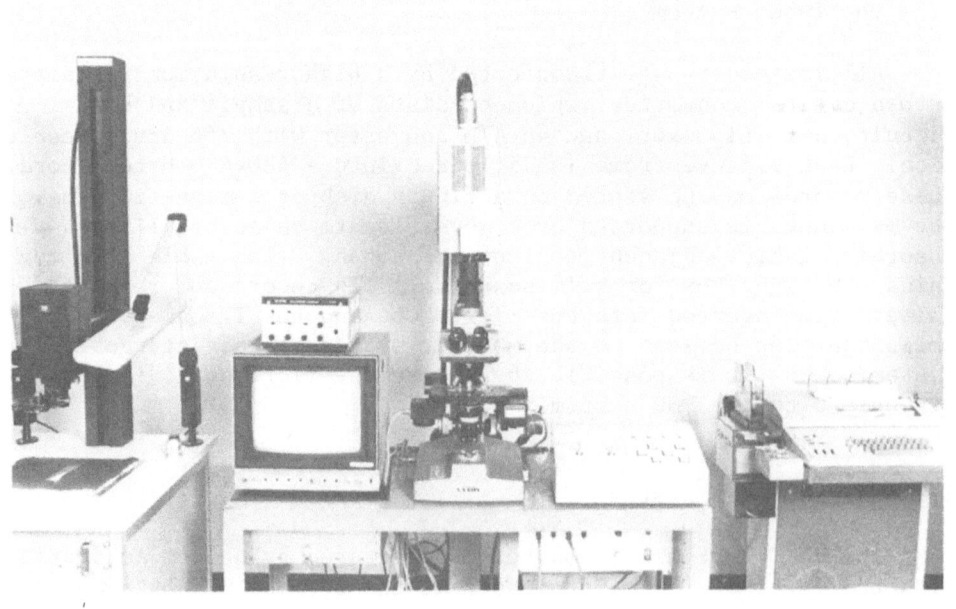

Fig. 8. The computerized microscope.

4.   DISCUSSION

The hardware of the two systems were supplied by several makers under the direction of the authors who then carried out basic calibration and testing. Both systems have passed their testing stages, and we are now developing several application programs. As was stated at the beginning, the complex system is intended to give service in a wide variety of medical applications (Table 1).

The application of highest priority is the analysis of vessel lumens from femoral angiograms, which is carried under a joint research activity with the cardiology department of our institute. Our interest is also in the following areas: electron microscopic picture analysis for three dimensional structure of aggregates of macromolecules, chromosome and cell classifications, stereologic measurements of microscopy images, x-ray film analysis.

A quite different application area is to use the image display as interactive data analysis facility. A comprehensive data manipulation tool based on pattern recognition techniques will be developed.

These applications have been carried out not only by the staff members of our joint research unit but also by medical consultants and basic researchers outside of our institutes who are interested in picture processing as joint research activities.

Table 1 - List of Ongoing Applications of Image Analysis
and Display at Tokyo Metropolitan Institute of Medical Science

Analysis	Femoral angiogram Muscle fine structure Muscle cell Electrophoresis Stereologic measurements Immuno assay
Recognition	Blood cell Chromosome classification Sister chromatid exchange Gastric Roentgenogram
Data analysis display	Dynamic clustering Catastrophy model

5.   ACKNOWLEDGEMENTS

The authors wish to thank the vendors who cooperated closely with us in developing the system hardware. We also thank those who cooperated closely when we developed the image analysis system: Dr. H. Yamazaki and Dr. T. Motomiya at TIMS and Dr. T. Soma at the Institute of Physics and Chemical Research.

6.   REFERENCES

Agrawala, A. K. (ed.), Machine Recognition of Patterns, New York IEEE Press (1977).

Kaminuma, T., Proceedings of the 9th Japan Image Engineering Conference, Tokyo, 1978 (in Japanese).

Onoe, M. and Takagi, M., "An Automated Microscope for Digital Image Processing, Part I: Hardware," in Digital Processing of Biomedical Images (Preston, K., Jr., and Onoe, M., eds.), University of Tokyo Press (1976), p. 17.

Onoe, M., Takagi, M., and Toshiro, T., "An Automated Microscope for Digital Image Processing, Part II: Software," in Digital Processing of Biomedical Images (Preston, K., Jr., and Onoe, M., eds.), University of Tokyo Press (1976), p. 17.

Preston, K., Jr., "Computer Processing of Biomedical Images," Computer (May 1976), p. 54.

CELLULAR COMPUTERS AND BIOMEDICAL IMAGE PROCESSING

S. R. Sternberg

Environmental Research Institute of Michigan

Ann Arbor, MI 48107, USA

1.   INTRODUCTION

The serial nature of conventional computers seems to restrict
or, at least, obscure their usefulness as image or picture processors. Since each computer instruction typically affects only
one or two pieces of data, manipulations of an entire image must
be accomplished through explicit iteration which costs heavily in
both time and conceptual distraction. The picture processing
approach we are investigating avoids these problems by defining a
set of operations which acts on the image as a whole. This provides an efficient conceptual framework for picture processing
tasks. In addition, the cellular nature of these operations allows
them to be implemented in computer hardware architectures employing
a high degree of parallelism thus eliminating the costs of iteration. To describe this type of picture processing computer we have
coined the term "cytocomputer" after the Greek "cyto" which means
"cell." A first generation cytocomputer has been developed at the
Environmental Research Institute of Michigan (ERIM).

Cytocomputer image processing operations are based on the concepts of cellular automata initially introduced by Ulam (1957) and
von Neumann (1961, 1966). Each cell or picture element of an image
is subjected to a sequence of time-discrete transformations, the
transformed value of a cell being determined by the initial values
of a finite group of cells composing its neighborhood. Each image
transformation is performed in an individual cytocomputer processing
element referred to as a processing stage. A cytocomputer consists
of a serial pipeline of programmable processing stages, where each
stage in the pipeline performs a single transformation on an entire

image. Pictures are entered into a cytocomputer in a line-scanned
format and progress through the pipeline of processing stages at a
real-time rate. Following an initial delay to fill the pipeline,
images can be processed at the same rate they are scanned.

2.   CELLULAR COMPUTERS

Computer image processing is the activity of retrieving the
relevant pictorially encoded information from an image. Conventional computers do not readily lend themselves to image processing.
Digital image manipulation by conventional computer is accomplished
only at a tremendous cost in time and conceptual distraction. Algorithm development for image processing is an alternating sequence
of inspired creative visualizations for the desired processed results and the formal procedures implementing the desired process on
a particular image processing system. Implementation of the visualized image manipulation by conventional computers requires frag-
mentation of the pictorial concept into information units matched
to the word-oriented capabilities of general purpose machines.
Conventional computer image processing could be broadly categorized
as manipulation of pixel states rather than pictorial content.

The image processing approach we are investigating differs
from conventional methods in that the basic manipulative informational unit is pictorial and deals with images as a whole.
Image processing is treated as a computation involving images as
variables in algebraic expressions. These expressions may combine several images through both logical and geometric relationships. Finally, highly efficient cellular computer architectures
are realizable for implementing the computation automatically.

2.1   The von Neumann Cellular Automaton

The concept of a cytocomputer arises out of the theory of
Cellular Automata. A cellular automaton can be roughly described
as a machine composed of a number of identical smaller machines
referred to as cells. Cellular automata theory was largely developed by von Neumann (1961). Von Neumann was concerned with the
behavioral aspects of both man-made automata and natural biological systems. He wished his theory to deal with the control, information, and logical aspects of systems composed of discrete
elements. One problem von Neumann posed and essentially solved
was the question of what kind of logical organization of discrete
elements is sufficient to allow an automaton to reproduce itself
(see von Neumann, 1966). Von Neumann envisioned a finite kinematic automaton, composed of interconnected parts, or elements,
which would be capable of manufacturing a copy of itself when

placed in a storeroom of unconnected parts. His model required him
to consider the kinetics of finding parts, cutting and fusing them
together and coordinating the entire self-reproduction process.
The kinetic system proved so complex that it was abandoned in favor
of a more manageable system which preserved von Neumann's primary
interest in machine behavior but eliminated many of the complexi-
ties of the kinematic systems. The new framework for examining
machine behavior was suggested to von Neumann by Stanislav Ulam
(1957). The framework is the concept of a cellular automaton.

A cellular system constitutes a "space" in which automaton
events can take place. Governing the events are simple rules
which precisely formulate what kind of action can occur in the
system. The cellular space consists of an infinite n-dimensional
Euclidean space together with a neighborhood relation which gives,
for each cell in the space, a list of cells which are its neighbors. A cellular system is specified by assigning to each cell a
"cell state" and a rule, called the "transition function," which
gives the state of a cell at discrete time t+1 as a function of its
own state and the state of its neighbors at time t. A special cell
state, called the "quiescent state," significes an "empty" or in-
active cell. The value zero is usually used to signify the quiescent state.

Once the initial configuration of cell states and the transition function are determined, the cellular automaton behaves in a
completely deterministic fashion. Von Neumann was able to find a
two-dimensional configuration of cell states and a transition rule
defined on a set of nearest neighbors such that, if this configuration were placed in a suitably large area of quiescent cells, it
would proceed upon repeated application of the transition function
to create an exact duplicate of itself by extending construction
"arms" out into the blank space. To make the self-reproducing
automaton non-trivial, von Neumann required it to contain within
its structure a Turing machine.

2.2 Cellular Automata and Image Processing

Cellular spaces and digital images share a common conceptual
framework. Each picture element of a digital image can be thought
of as a cell in a given state. If we define a neighborhood relationship and a cell transition function on a digital image, then
application of the transition function will cause the configuration
of cell states forming the image to be modified or transformed
into new configurations. Of particular interest is the question
of whether there exist neighborhood relationships and transition
functions which will cause images to be transformed in predictable
and useful manners.

The studies of cellular automata and cellular digital image processing differ in several aspects. Whereas in cellular automata studies the search is generally for initial configurations which exhibit desirable behavior patterns when a single transition function is repeatedly applied within the cellular space, the emphasis in cellular digital image processing is to find a sequence of transition functions, generally not identical, which cause a given initial configuration to behave in a predictable fashion. The sequence of transition functions constitutes an image processing program or algorithm.

A physical manifestation of a finite cellular automaton is the two-dimensional modular array illustrated in Figure 1. The elements of the modular array are cells or modules. Each cell contains a memory register for holding the current state of the cell. The two-dimensional configuration of cell states constitutes a digital image. Each cell is directly connected to the set of nearest neighbor cells in the array forming the neighborhood relationship.

Cells of the modular array contain a neighborhood logic module for computing cell transitions. Cell transition instructions are programmed in parallel into each cell's neighborhood logic module

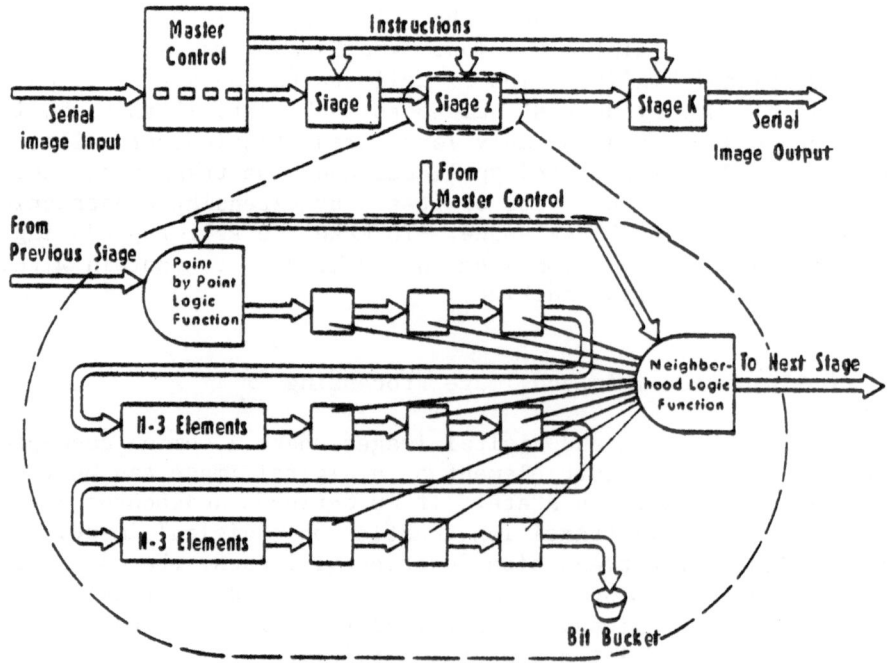

Fig. 1. Modular array image processor.

by the Master Control which stores the complete sequence of cell
transition instructions constituting the algorithm. Input to each
cell's logic module are the states of each cell in the neighborhood.
All array cell transitions are effected simultaneously. The output of the cell neighborhood logic module is stored in the cell
state register. A complete image transforming algorithm is a sequence of such neighborhood transitions each applied to all cells
in parallel. It is executed by alternately having the Master Control place cell transition instructions (in parallel) in the neighborhood logic modules and having the logic modules execute these
instructions (in parallel) to perform the cell state neighborhood
transformations. A number of modular array image processors have
been proposed or constructed and are reviewed by Preston et al.
(1979).

2.3 Pipeline Image Processor

Although the modular array architecture illustrated in Figure 1 represents the most obvious implementation of a cellular
digital image processor, there exist a variety of machine architectures which will transform images in a systematic cellular fashion.
The pipeline processor represents an alternative approach to a
cellular digital image processor. The basic replicated functional
element of the pipeline cellular digital image processor is called
the "stage" (Figure 2). The function of the stage is to execute
a single transition function over an entire digital image. Input
to a stage is a sequence of array cell states, arranged in raster
scan format.

Shift register delays within the stage store contiguous scan
lines while taps on the shift register delays extract in parallel
the neighborhood states which constitute the input of the Neighborhood Logic Module. At each discrete time interval, a new cell
state is clocked into the stage. Simultaneously, the contents of
all shift register delay elements are shifted by one element. The
neighborhood configuration extracted by the shift register taps
is thus sequentially scanned across the input array in raster scan
format. Neighborhood logic transformations are computed within
the basic clock period allowing the output of a stage to occur at
the same rate as its input.

A pipeline cellular digital image processor consists of a
serial concatenation of shift register stages and its Master Control. The Master Control programs each sequential step of the
image processing algorithm into the serially configured stages.
Raster scanned images which are fed through the pipeline are
processed one algorithm step per stage.

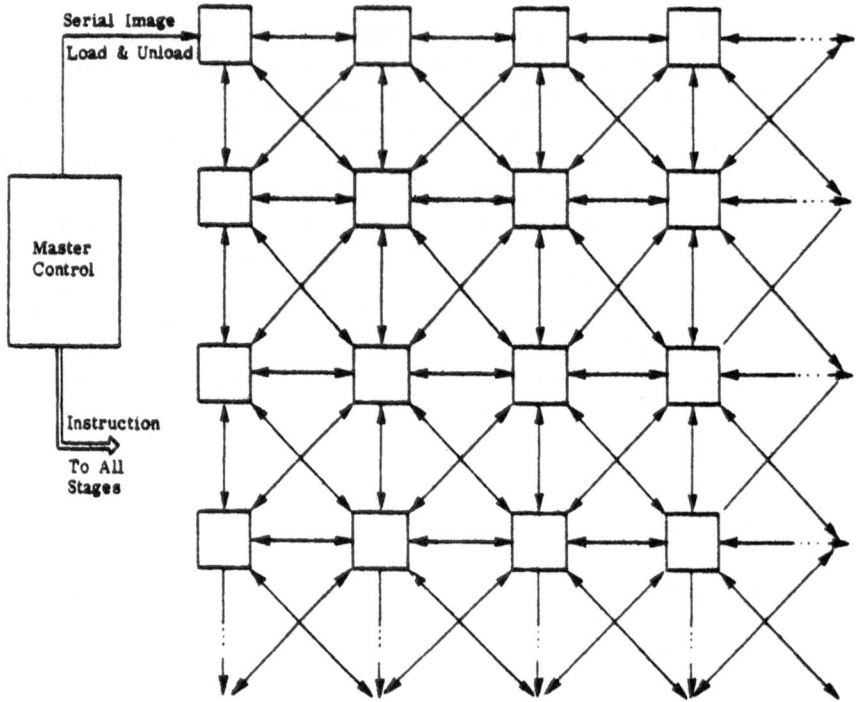

Fig. 2. Pipeline image processor showing one stage in detail.

3.   CELLULAR COMPUTERS AND IMAGE PROCESSING RESEARCH

Parallel processing of digital images by large numbers of identical electronic modules is a particularly timely idea. On one hand, we have a theory of cellular automata which states that a cellular array can simulate a Turing machine; hence, it is a computer in the formal sense. On the other hand, we have the recognized requirement for computer architectures employing high degrees of parallelism for achieving the kind of computational power needed for near real-time processing of images containing millions of bits of data. The application of cellular computer methods to digital image processing is the foreseen solution to these requirements.

Problems of fabricating large numbers of complex electronic cells and cell interconnections have been largely overcome. In fact, computer systems analysts broadly agree that major advances in computer architectures will come about as the result of the modularization of parallel data flows and the realization of an integrated circuit module to be incorporated in large numbers in every computer mainframe. As far as digital image processing is concerned, the problems of hardware realization are well in hand.

3.1 Programming the Cellular Computer

Understanding how to program a cellular array for image processing is another matter. As a programmer's well developed faculty for computer programming is largely the result of his/her innate ability to manipulate strings of symbols, i.e., language, we are well suited for the task of modeling complex systems in conventional serial processing computers which require the formulation of a problem into linear strings of operations. But picture processing on a cellular computer is another matter. Since each instruction typically affects millions of bits of data, the problem of visualizing the anticipated result of a computation is of considerable importance.

A significant portion of our research effort at ERIM in applying cellular computers to image processing is in developing a formal language for parallel picture processing. An early achievement in this direction has been the development of C3PL (Cytocomputer Parallel Picture Processing Language) for programming the ERIM Cytocomputer. C3PL is a command interpreter. Its major significance is in permitting the interactive manipulation of digital images by cytocomputer parallel neighborhood operations. Its principal weakness is that it is not possible in general to conceptually formulate any picture processing operation conveniently in C3PL, and similarly it is not immediately apparent from inspection of C3PL programs what the program is doing.

The apparent ease with which humans interpret pictures is in marked contrast to their understanding of how they do it. Not so in other areas of mental consideration where the ability to perform information manipulation is accompanied in large part by the ability to describe the manipulation process abstractly. For this reason programming for cellular image processing is largely an interactive affair. The results of a particular image manipulation must be made immediately available for inspection in order to provide on-line feedback to reinforce or inhibit the programmer's fledgling conceptualizations as to what constitutes a language representation of a picture processing task.

3.2 Research Status

Image processing is applied science and research in cellular image processing ought to be guided by a principle of applicability of developed results to problems in application areas whose solutions represent significant progress in that respective area. Our research is application directed, although many aspects of the research are of an academic nature. The validity of a theory in image processing is thought to be demonstrable only when the application of that theory leads to a successful employment in terms of the processing of a large number of images from real world sources.

In summary, our research in cellular methods of digital image processing concentrates in three areas. First, the investigation and formalization of cellular computer architectures for parallel processing; secondly, the development of languages for parallel image processing; and thirdly, the interactive application to real world images. Our emphasis thus far has been on the application of cellular image processing techniques to biomedical image processing. The range of types of images and image processing requirements being as broad in the biomedical sciences as can be found anywhere in the applied world.

4. BIOMEDICAL APPLICATIONS OF CELLULAR DIGITAL IMAGE PROCESSING

The ERIM Cytocomputer, a pipeline cellular image processor, consists of a total of 113 stages and processes a digital image as large as 1024x1024 pixels at a rate of 1.6 megapixels per second. Demonstrations of the feasibility of employing cellular digital image processing to biomedical image analysis have been concentrated on three different application areas: (1) heart disease, (2) tissue growth and development, and (3) genetic mutation. Experiments have been performed in close collaboration with J. H. Thrall, T. G. Connelly, and J. V. Neel in the University of Michigan Medical School.

4.1 Coronary Artery Disease

Imaging of the cardiovascular system has become an essential part in establishing the diagnosis of many cardiovascular diseases especially coronary artery disease and in assessing the degree of functional impairment. Quantification of image parameters has been a well recognized goal in cardiac diagnosis for over two decades. Based on results of tedious hand calculations and, more recently, semi-automated analyses there is now an overwhelming consensus that quantification not only increases diagnostic accuracy but also provides significant data not obtainable from qualitative image analysis alone.

The first step in quantitative image processing is contour mapping to define the limits of cardiac structures, most importantly, the left ventricle. This step has remained a major stumbling block in meeting the requirements for an ideal quantitative analysis system. A cytocomputer processing algorithm has been developed for contour mapping of the left ventricle in nuclear ventriculograms. The algorithms have been run on both phantom and clinical data. Correlation of the cytocomputer generated outlines and the geometric measurement of the phantoms has been excellent. Likewise, comparison of cytocomputer determined left ventricular contours and results of conventional analysis has been excellent in the clinical studies.

Color Plate 1* shows a typical nuclear ventriculogram time-gated bloodpool study consisting of 16 64x64 pixel subimages. Sub-images are sequenced from end diastole (upper left) through end systole to end diastole (lower right). Abnormal cardiac function causes virtually indistinguishable left ventricle contours. These contours are enhanced and displayed as pseudocolor images (Color Plate 2). Treating the intensities of the images of Color Plate 1 as heights, a spherical reference image having a 12-pixel diameter is "rolled" over the brightness-height surface of the composited frame by iterative neighborhood transforms. The sphere, being wider than the narrow "valleys" between cardiac substructures, rolls along the peaks of the substructures. The differences be-tween the heights of the original frame (Color Plate 1) and the heights of the sphere-normalized frame are shown in Color Plate 2.

The noisy ventricular edges shown in Color Plate 2 are smoothed and a best estimate of the left ventricle outline is superimposed on the data of Color Plate 2. Smoothing is accomplished by shifting the individual subimages to form sixteen subimage composites of three subimages each, the subimages taken at time intervals t-1, t and t+1, where t = 0, ..., 15 modulo 15. Processing of the six-teen three-subimage-composites is performed in parallel by iterative neighborhood transforms. The result is shown in Color Plate 3. Finally, left ventricle outlines obtained from Color Plate 3 are superimposed on the original frame. The automatically extracted ventricular contours permit a cardiologist to diagnose ventribular function or to quantify ventricle performance. See Color Plate 4.

4.2 Tissue Growth and Development

One of the major areas of interest in growth and development is the method by which cells determine their position relative to other cells. This determination of position is important for the differentiation of the cells and for morphogenesis as well. Several theories have been proposed to explain growth pattern formation and several mathematical formalizations of growth pattern formation models have been put forth. A problem inherent in all investiga-tions of this sort is that of obtaining sufficient data to estimate the model's parameters and validate its predictive performance.

---

*The color plates will be found following page 144.

Data are now obtained from manual counts of labeled dividing
cells and of overall cell density. This is tedious and time consuming. The limited amount of data thus obtained, and the inherent bias of manual counts weaken both the statistical analysis
and the interpretation of results. Reported experiments are difficult to duplicate due to the difficulty faced in obtaining detailed
and statistically meaningful cell density and mitotic index counts
in complete structures using conventional methods.

A cytocomputer algorithm has been developed and evaluated for
automatically counting and discriminating unlabeled and labeled
dividing cells in newt limb-bud sections and computing densities
of cells and mitotic activity. Comparison with manually analyzed
sections has demonstrated the consistency and accuracy of the
automatic processing. The speed of limb-bud section analysis is
presently permitting investigators to consider the three-dimensional
aspects of the growth and morphogenesis of the limb and the relationship of cell dynamics within the limb to these processes.

Color Plate 5 shows a stained section of newt limb-bud tissue
at a resolution of 1024x1024 pixels. Cell nuclei have been automatically counted and cued by a white dot with a sequence of cytocomputer neighborhood operations (Color Plate 6). Processing time
is approximately one second. Color Plate 7 illustrates a transformation stage prior to segmenting overlapping nuclei. Nuclei
centers are topologically distinguishable from annular rings which
surround them. Another approach to segmenting overlapping nuclei
is shown in Color Plate 8. Nuclear shapes are decomposed into
unions of convex blobs which are color coded by size (green, blue,
and white). These colors and the spatial connectivity patterns
formed by the blobs infer syntactical rules for segmenting adjoining
or overlapping nuclei. Yellow and red are used to color code the
connective links between blobs. (See Color Plage 9.)

Color Plate 10 shows a composite of newt limb-bud tissue sections. The left hand image is an autoradiograph which displays
radioactively-tagged, recently-formed nuclei as dark spots. The
right hand image is the same tissue section stained to enhance all
cell nuclei. Low resolution images were digitized on a video microscope and then cytocomputer processed. The results of cytocomputer
processing are shown in pseudocolor in Color Plate 11. Here
density maps of both recently developed nuclei (left) and total
nuclei (right) display time-differential mapping of morphological
limb-bud development. Individually processed tissue sections are
then combined into a stereological growth model of newt limb-bud
regeneration.

4.3 Genetic Mutagen Studies

There is a significant concern today about man's exposure to
environmental elements, especially to those agents which may be
carcinogens or mutagens. As a step in quantifying the impact of
mutagenic agents on human populations, techniques are being developed for the detection of mutations and for estimating the mutation rate in sample populations. Electrophoretic methods are
utilized to detect protein structural alterations which reflect
changes in genetic material. Recently, two-dimensional electrophoresis systems in which proteins are separated according to
charge in one direction and on the bases of subunit molecular
weight in the second direction have been developed. The significant practical advantage of the 2-D gel system is that many proteins are identified on a single gel.

Gel protein patterns are complex and direct visual inspection
for genetic variants is almost impossible. Estimating the human
mutation rate requires the sampling of thousands of mother-father-
child triads. Automated gel reading and analysis is thus an absolute necessity. An important aspect of current research is the
development of cytocomputer techniques to detect genetic variants
in 2-D gels. A composite of two digitized 2-D gel autoradiographs
of human hair follicle proteins is shown in Color Plate 12. Gels
are from two different subjects but display a marked similarity.
Differences constitute genetically relevant information. Color
Plate 13 is a pseudocolor composite of details taken from these
2-D gel patterns. Obvious image differences arise from both
genetic variations (missing spots) and inhomogeneities in the gel
electrophoresis processing (variable background intensities).

Cytocomputer processing consists, first, of rolling a sphere
whose diameter is greater than the spot widths under the height-
brightness surface of the gel image to remove the spatially distributed background level. This does not modify the spot shapes
and permits multilevel quantization across the entire gel pattern
as shown in Color Plate 14. Next spot boundaries are detected by
iterative neighborhood transforms. The spot boundaries are in-
ferred by topological properties and the relationships of individually quantized levels as shown in pseudocolor in Color Plate
15. Finally, protein spot boundaries are superimposed upon the
composite gel detail. (See Color Plate 16.) Missing spots infer
an inactivated gene or a genetic variant, hence, detection of very
low intensity spots is a critical step in gel image processing.

5.   REFERENCES

Preston, K., Duff, M. J. B., Levialdi, S., Norgren, P. E., and
      Toriwaki, J-i., "Basics of Cellular Logic with Some Applica-
      tions in Medical Image Processing," Proc. IEEE 67(5):826-856
      (1979).

Ulam, S. M., "On Some New Possibilities in the Organization and
      Use of Computing Machines," IBM Research Report RC68 (May
      1957).

Von Neumann, J., "Design of Computers, Theory of Automata and
      Numerical Analysis," in Vol. 5 of Collected Works (Taub,
      A. H., ed.), New York, Macmillan (1961).

Von Neumann, J., Theory of Self-Reproducing Automata (Burks, A.,
      ed.), Urbana, University of Illinois Press (1966).

ULTRA HIGH-SPEED RECONSTRUCTION PROCESSORS FOR X-RAY COMPUTED

TOMOGRAPHY OF THE HEART AND CIRCULATION

B. K. Gilbert, R. A. Robb, L. M. Krueger

Biodynamics Research Unit

Mayo Clinic, Rochester, Minnesota   USA

1.   INTRODUCTION

Present commercial x-ray computed tomography (CT) scanners are capable of producing single cross section images of the head, thorax, or abdomen which exhibit both high spatial and halftone resolution as described by Brooks and DiChiro (1976). However, the operational flexibility of these devices is restricted by their inability to record sufficient x-ray transmission data to reconstruct more than a single 5-15mm thick cross section of tissue during one scan and the relatively long duration (2-60s) for each scan. This renders such systems incapable of "freezing" the motion of rapidly moving structures, e.g., the beating heart and breathing lungs. Thus, the single cross sections generated in this manner cannot provide clinical diagnostic or biomedical research data concerning the true anatomic shape and dynamic changes in the structural dimensions of several important organ systems.

Development of an advanced cylindrical scanning multiaxial tomography unit called the "Dynamic Spatial Reconstructor" (DSR) has been undertaken which will be capable of imaging the thoracic and abdominal organs with high temporal and spatial resolution. The DSR is designed to scan over a span of 24cm in the cephalocaudal (z-axis) dimension of the subject, and can repeat the complete scan procedure at 16ms intervals, allowing dynamic changes in shape and dimensions over the entire 3-dimensional anatomic extent of the intact beating heart and breathing lungs and the vascular beds of otherwise stationary organs such as the kidney to be reconstructed at a repetition rate of sixty per second. A complete description is given by Ritman et al. (1978). Figure 1 is a montage

(a)

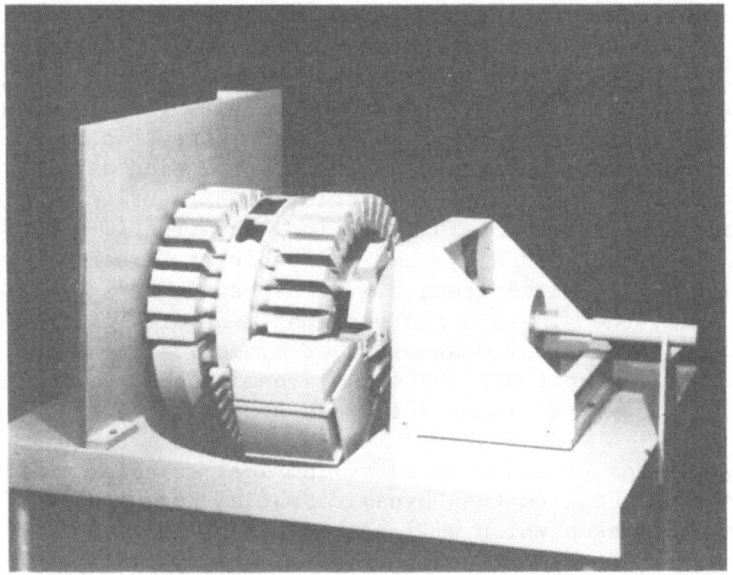

(b)

Fig. 1. One-fifth scale model of the Dynamic Spatial Reconstructor. On facing page: (a) Patient is positioned along rotational axis of gantry on a cantilevered radiolucent table through opening in interface wall. (b) Small boxes positioned around gantry contain electronics for image intensifier/video chains, each in a cylindrical metal tube. Above: (c) Large semi-annular cover houses high

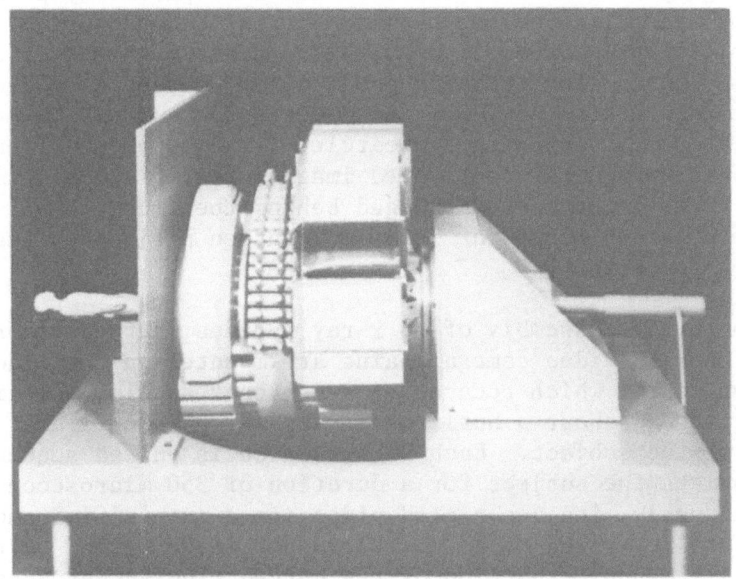

(c)

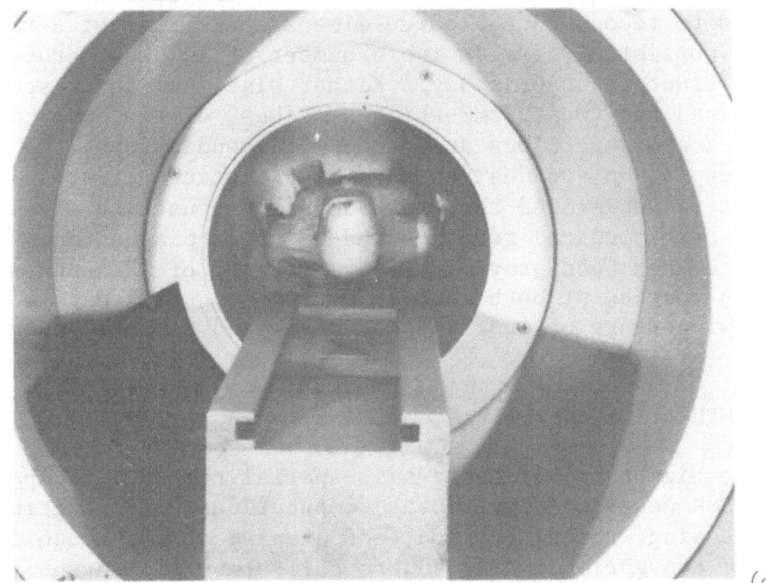

(d)

voltage power bus and high voltage modulators for x-ray tubes.
Rear triangular support structure contains support beating and
high voltage slip rings; note low voltage slip rings for video and
control signals. (d) Close-up of patient access port, tunnel, and
table support.

of photographs of a scale model of the multiple x-ray source and
detector assemblies of the Dynamic Spatial Reconstructor, currently
undergoing fabrication. The system consists of a set of 28 independently controlled rotating-anode x-ray sources arranged around
a semicircle of radius 143cm and a 30cm wide curved fluorescent
screen around an opposing semicircle of radius 58cm. Twenty-eight
sets of independently controlled image intensifiers and image
isocon video cameras are arranged behind the fluorescent screen to
produce 28 video images of 30x30cm roentgen projection images produced on the screen.

The entire assembly of 28 x-ray sources, curved fluorescent
screen, and 28 video camera chains are mounted on a cylindrical
aluminum gantry which rotates at a constant angular velocity of
90° per second about a horizontal axis which is co-linear with the
z-axis of the subject. Each x-ray source is pulsed sequentially
to irradiate the subject for a duration of 350 microseconds.
Simultaneously, its associated video image intensifier and video
camera are activated so as to record the image formed on the dia-
metrically opposite portion of the curved fluorescent screen.
Each scanning sequence provides a sufficient number of x-ray projections to reconstruct a three-dimensional array of x-ray density
values, consisting of a variable number of adjacent cross sections
of the structure under study. Either sixty 4mm-thick sections,
one hundred and twenty 2mm-thick sections, or two hundred and forty
1mm-thick cross sections are produced depending upon the required
axial resolution and desired photon statistics per cross section.
Since it is envisioned that the DSR will be used to investigate a
variety of biomedical research and clinical diagnostic protocols,
the system has been provided with a variety of scanning modes
allowing imaging of both rapidly moving and stationary structures
with a wide range of spatial, contrast, and temporal resolution.

2.   COMPUTATION AND DISPLAY

The high temporal and z-axis spatial resolution capabilities
of the DSR generate significant computational difficulties in the
preprocessing, reconstruction, and display of the image data. For
example, conventional CT scanners collect sufficient projection
data within five to twenty seconds to reconstruct a single cross
section. By comparison, proposed DSR scanning protocols for the
diagnosis of the presence and extent of coronary artery disease
require the recording of approximately five to ten seconds (i.e.,
five to ten complete cardiac cycles) of projection data from each
subject, creating a computational load of many thousands of 127x127
element cross sections from each study.

In the first phase of operation of the DSR all data processing will be carried out by a general purpose "maxi-mini" computer c coupled to a floating-point array-processor to speed the calculation of the innermost loops of the reconstruction algorithm and a special-purpose high-speed digitization, data handling, and pre-processing computer called the Mayo-TRW High Speed Interface (HSI), which itself is augmented with 2 megabytes of 100ns (cycle time) static random access memory (RAM) as described by Gilbert et al. (1976).

The HSI will provide, under computer control, high-speed digitization (14 megasamples/second) of the video projection images recorded on multiple video disc platters employed for intermediate analog storage of the DSR camera outputs. The HSI will also provide high-speed storage and retrieval of these data with or without application of various digital data compression/decompression algorithms.

The array processor attached to the maxi-mini computer is a microprogrammable device featuring a basic cycle time of 167ns, five separate memories, and a parallel "pipeline" multiplier and adder, all of which provide array processing capabilities in floating point arithmetic. In particular, use of the device is expected to permit fan-beam convolution reconstruction of single cross sections in less than 10 seconds. This capability is extremely important, since a single DSR cylindrical scan of even one second duration at the minimum z-axis resolution can provide projection data for 3,600 cross sections (60 profiles at 60/s). About 4 minutes would be required on the minicomputer, operating unassisted, to complete a fan-beam reconstruction of a single 128x128 cross section. Therefore, approximately 300 hours would be required for the minicomputer to reconstruct the 3,600 cross sections covering the entire anatomic extent of the thorax for each 1/60 second interval recorded during a 1 second DSR scan period. However, in conjunction with the array processor, all 3,600 cross sections could be reconstructed in about 10 hours, or a factor of 30 times faster using the minicomputer-array processor combination than would be possible using the minicomputer alone. Since most DSR scans will be of five to ten seconds duration and generate a minimum of 18,000 to 36,000 cross sections, such a performance improvement assumes even greater significance.

Practical use of the DSR in the initial development and evaluation phases will dictate that a limited number of cross sections be reconstructed for any given procedure. For example, a three-dimensional volume of the chest consisting of 60 cross sections recorded in one 16ms interval of operation of the DSR, would require about 10 minutes to compute with the minicomputer-array-processor system. Similarly, some "functional" studies consisting of limited

numbers (e.g., 10 to 15) of these 60-cross-section volumes determined throughout individual cardiac and respiratory cycles would
be possible. Optimal routine application of the DSR for high
temporal resolution studies of patients and/or animals must, however, await development of the ultra high-speed (i.e., 100 cross
sections/sec) special purpose reconstruction capabilities, as
described below.

2.1 The Cone Beam Geometry

Commercially available x-ray tubes are in effect point sources
of x-rays, radiating energy in a right elliptical conical beam.
The earliest reconstruction algorithms could not account for the
geometry of conically divergent beams, resulting in the constraints
that all rays at each angle of view had to be parallel to one
another and entirely within the plane of the cross section to be
reconstructed. X-ray sources of early CT machines were thus
physically collimated to a pencil beam which mechanically traversed
the patient, with transmission data collected via a single collimated x-ray crystal detector. Recent algorithm developments permit
a divergent x-ray beam to traverse the slab of tissue to be reconstructed, i.e., a "fan" beam. Commercial scanners now physically
collimate the beam to a planar fan and detect the transmitted
radiation with a one-dimensional array of x-ray detectors. After
several years of intensive effort by a number of researchers, no
analytic closed form algorithm has been theoretically derived
assuming a cone beam of x-rays. As a result, x-ray scanners must
either collimate the source beam to a fan and forego the oppor-
tunity to collect projection data simultaneously for many adjacent
cross sections, or assume that the conical beam can be approximated
by a "stack" of adjacent fan beams in the axial direction (Robb et
al., 1979). Though incorrect in the absolute sense, the assumption
of adjacent fan beams does not appear to introduce intolerable
errors, since for present x-ray sources the divergence angle of the
cone beam in the axial direction does not exceed 3°. Thus, the
assumption of parallel axial fan beams for cone beam geometry is
liekly to be continued for the near future, under the premise that
the availability of three-dimensional images (volumes) containing
second order geometric distortions is more useful than single two-
dimensional images.

2.2 Choice of a Suitable Reconstruction Algorithm

The development of a high capacity computer to operate in
support of the DSR required an examination of candidate reconstruction algorithms potentially suitable for execution by special purpose processors. Experience has indicated that the family of

filtered back-projection algorithms produces images of quite satisfactory spatial and gray scale resolution and executes rapidly on
general purpose computers. The high execution speed and satisfactory image quality of the fan beam algorithms disclosed by Herman
et al. has motivated our attempts to implement them in a special
purpose processor. Based upon studies of the numerical characteristics of the fan beam filtered back-projection algorithm, a tentative design has been established for a reconstruction processor
capable of exploiting the computational parallelism inherent in
the algorithm. The first stage of this algorithm requires the one-
dimensional linear filtration of all x-ray projections with a
single, preselected filter function. In the second portion of the
algorithm, the back-projection of the filtered projections into
the image space yielding the density value of each cross-sectional
image element (pixel) is determined by selecting one sample (ray
sum) from each of the filtered projections according to a pre-
specified algorithm, followed by sequential addition of all
selected ray sums to create the density value of that pixel.

2.3 Filtration Methods

The filtration of the individual projections can be carried
out by any viable linear filtration method. For computed tomography
applications, direct convolution filtration has been the most
heavily employed method, in contradistinction to the current usage
in other disciplines of the classical Fast Fourier Transform (FFT)
to implement a so-called "FFT-Fast Convulution" algorithm. With
the exception of data vectors of 32 elements or fewer, the number
of multiplication and addition operations required for Fourier
filtration is considerably smaller than for direct convolution
approaches. Historically, the relative cost of digital arithmetic
operations executed on general-purpose computers has been so great
that the reduction in absolute numbers of arithmetic operations
achieved by the FFT-Fast Convolution method could result in a significant reduction in computational cost. Nonetheless, several
inherent characteristics of the classical FFT algorithms present
their own unique set of complications. FFT convolution of real
vectors requires execution of numerical operations in the domain
of complex numbers; further, because these computations require
use of trigometric functions and thereby of irrational numbers,
the availability of floating point arithmetic hardware is desirable
(though not mandatory) to minimize computational roundoff error.
A final restriction in the use of classical FFT methods arises from
the difficulty of employing digital pipeline processing techniques
for the individual substeps of the FFT operation, thereby dis-
couraging the design of hardwired FFT processors which truly exploit
all the arithmetic parallelism apparent in the algorithm. (See
Groginsky and Works, 1970.)

Recently, the theory of linear transformations has been significantly expanded through the development of number-theoretic
and finite field transforms by such workers as Ramachandran and
Lakshminarayanan (1971). These transforms are a class of linear
transformations defined to exist only on carefully selected finite
fields of numbers, e.g., the subspace of real rational (integer)
values. Although these new transforms require a much lower absolute number of arithmetic operations even in comparison to the
classical FFT, they appear to require significantly greater numbers
of logical operations and memory references, as well as very complicated control sequences to assure correct generation of intermediate
results. Though the finite field transforms appear marginally
faster than the FFT when executed on general purpose computers,
they seem less well suited to implementation on special purpose
processors, for which the cost of arithmetic is falling rapidly.

Conversely, direct convolution of real vector operands of
fixed precision can be carried out entirely within the domain of
real integers, obviating the requirement for floating point hardware configurations. The increasing availability of high-speed
large scale integrated circuit (LSI) fixed precision arithmetic
components of moderate cost, the conceptual simplicity of direct
convolution, the ease of developing both LSI device and system
architectures capable of carrying out the necessary arithmetic
operations in a pipelined manner, the freedom from large design
commitments to control logic, and the straightforward design,
fabrication, and testing of appropriate hardware, makes physical
implementation of the direct convolution approach appear both
feasible and practical.

2.4   Reconstruction Processor Design

Figure 2 is a design for a high capacity reconstruction processor optimized for the fan beam filtered back-projection algorithm. The upper portion of Figure 2 demonstrates schematically
the operation of a direct convolution filter. The digitized projections, each composed of I distinct samples corrected for camera
signal bias, normalized to incident x-ray exposure levels, and converted into density values, are supplied to the uppermost processor section which carries out the filtration. The blocks designated H_k represent a series of digital registers, each connected
to one input of a dedicated high-speed multiplication unit. Data
samples are entered consecutively into register H_1, and are shifted
into each successive register H_2, H_3, etc., and applied to the inputs of their associated multipliers. When the M or fewer non-zero
products thus generated during each clock cycle T_c are summed during
the next time interval T_c (by means of a parallel adder of addition
tree), one sample of the *convolved* projection is formed. As each

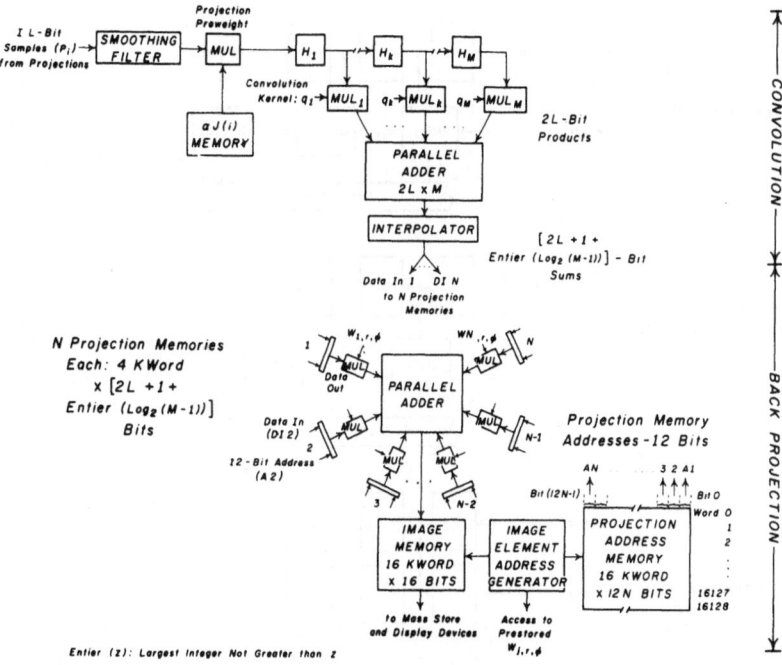

Fig. 2. Schematic of digital hardware implementation of fan beam convolution reconstruction algorithm, employing direct convolution reconstruction filtration and a memory controlled parallel processing back projection method.

projection is advanced through the series of registers, the entire convolution will be carried out within (I+M-1) clock durations T_C.

The design for the linear filtration processor can be simplified considerably by exploiting specific numerical characteristics of the filter function itself. The upper panel of Figure 3 depicts a subsection of the detailed architecture of a linear filtration processor used in conjunction with sampled data filter functions of arbitrary shape (represented graphically in the inset view in the upper right). If, however, it can be predetermined that alternate elements of the filter kernel are identically zero, i.e., if the filter function is so-called "half-band" (middle panel), so that the number of multipliers required to implement the direct filter can be reduced by nearly a factor of two, although the registers associated with the deleted multipliers must be retained. The use of half-band filters in computed tomography is not uncommon, as exemplified by the sampled data version of the filter kernel originally proposed by Ramachandran and Lakshminarayanan (1971) for their parallel-beam filtered back-projection reconstruction algorithm.

The lowermost panel of Figure 3 demonstrates an additional saving in hardware which can be achieved if the filter function is

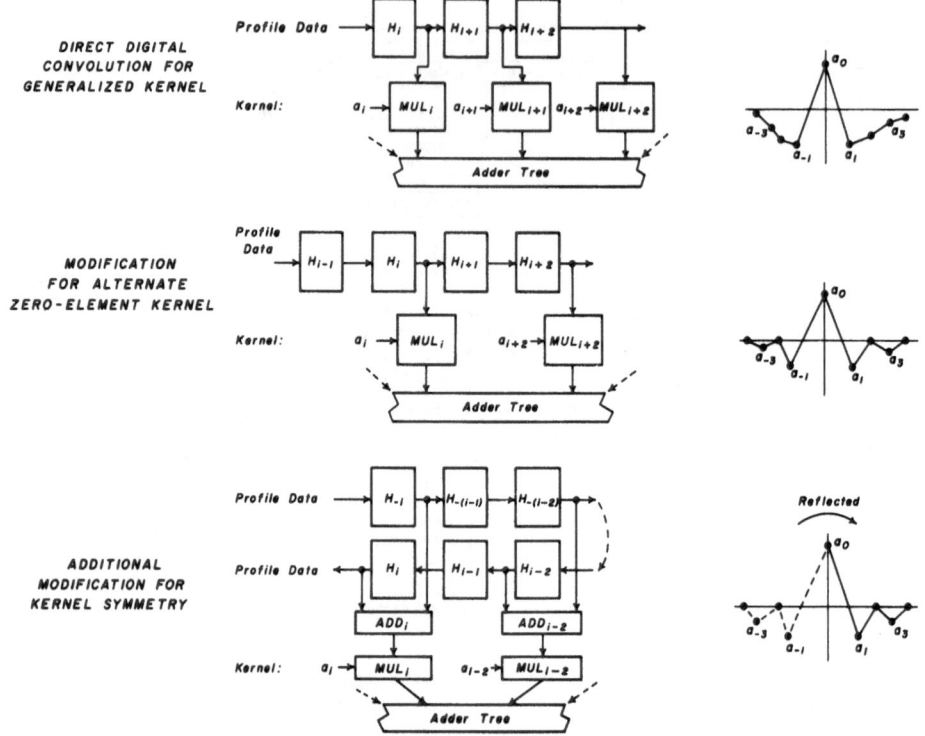

Fig. 3. Modifications of direct digital convolution processor to
account for special characteristics of filter function. Upper
panel: Selection of generalized filter for kernel elements are
zero, multipliers associated with null elements may be removed.
Lower panel: If kernel is an even function, it may be "folded"
about ordinate, halving required number of multipliers.

even, i.e., symmetric about the ordinate, in the spatial domain.
The number of physical multipliers may then be nearly halved by
reflecting the kernel about the ordinate; for example, since
filter samples $a_{-i} = a_i$, elements of the data vector which multiply
a_{-i} and a_i during each clock cycle are first summed, and thereafter
multiplied by the filter element a_i. Only the physical placement,
not the number, of adders is altered by the procedure, but the
requisite number of multipliers is nearly halved. That this approach
can always be invoked is assured by the requirement that a linear
phase response be maintained by the filter function to prevent phase
distortion of the filtered projections, and thereby of the complete
reconstructions. Bracewell (1965) has shown that linear phase response is guaranteed only if the filter function is pure real in
the spatial frequency domain, and therefore an even function, i.e.,
symmetric about the ordinate, in the spatial domain.

Additional simplifications of a direct convolution processor can be identified by exploiting techniques recently developed by Kenue and Greenleaf (1979) for converting an arbitrary convolution kernel into one in which the central value is an arbitrary fixed precision number, but all other values of the kernel are simple binary multiples of one another. The conversion of arbitrary filter weights into binary numbers makes it possible to execute the filtration operation primarily by shift and add operations, nearly obviating the requirement for execution of multiplications. Several of the computational savings inherent in these binary kernels can be exploited on a commercially available general or special purpose computer whose architecture is already established. However, greater improvements in throughput and cost-performance can be gained if new filter processor designs are developed and incorporated into special purpose high-speed reconstruction processors. Based on the results of Kenue's studies, we have developed a new design for a direct convolution processor which will permit rapid execution of a variety of binary-weighted kernels.

2.5 Back-Projection Implementations

Two significantly different physical implementations of back projection, referred to as "projection-driven" and "pixel-driven" back-projection, have been proposed. In the case of projection-driven back-projection, commonly implemented in high level languages for standard general purpose computers, the jth projection is filtered, its contribution to every pixel in the cross section is computed, the projection is discarded and replaced with the $(j+1)^{th}$ projection, and the procedure is repeated. That is, the ray sums for each angle of view are back projected for all image pixels before proceeding to the next angle of view (projection). The linearity property of the back projection operation allows the contribution of each projection to be computed independently and added to the running sums of contributions of all other projections. Sequential processing of projections conserves computer memory, since the data area allocated to the program need only be large enough to accommodate the N^2 image pixels of an NxN image and a single projection at any stage of the procedure. The ray sum contributions to each pixel are added to the accumulated contributions of previously processed projections by reading the contents of the memory location assigned to that pixel, adding the contribution from the projection currently undergoing back-projection, and writing the updated summation back into the same location within the image memory. Back-projection of M projections to form an image by this procedure requires at least $(M-1)N^2$ memory read operations and MN^2 memory write operations, i.e., $(2M-1)N^2$ memory references.

Pixel-driven back-projection, under development in our laboratories, is a parallel processing approach in which all filtered projections must be available continuously throughout the entire duration of the image formation. For this procedure, all the projection ray sums, one from each angle of view, whose rays intersect a single image pixel are back-projected before proceeding to the next pixel. Such an implementation executes the reconstruction algorithm in a highly parallel fashion because the contributing ray sums from all projections are indexed simultaneously and then summed simultaneously with the aid of one or more forms of non-sequential addition techniques, e.g., a rank of Wallace tree adders. Each pixel is thus completely back-projected in a single set of arithmetic operations and not readdressed thereafter.

The lower portion of Figure 2 is a schematic of a hardware processor designed to execute pixel-driven back-projection, assuming that N projections will be back-projected simultaneously. The filtered projections, in this example interpolated to a maximum of 4096 ray sums, are stored in a group of "projection memories" prior to the onset of back-projection. Each projection memory contribution is created by reading the contents of one data word location, multiplying this value by an appropriate "back-projection weight," as specified by Herman et al. (1976), and applying the weighted data values to the inputs of a high-speed multiple operand parallel adder which forms the required image pixel density sum. The correct data values from each projection for every pixel are specified by a unique set of N addresses which index, or point to, the requisite ray sums within all of the N projection memories. Thus, for a 127x127 cross-sectional reconstruction, there are 16129 unique sets of N 12-bit addresses (each 12-bit address can access any of the 4096 possible ray sums within a projection) which may be precalculated and prestored in a hardware subunit of the processor termed a "projection address memory." As diagramed in Figure 2, reading of a single 12-bit word in this projection address memory results indirectly in the reading of the necessary ray sum from each of the projection memories to generate one image pixel. With the availability of a multiple operand adder, e.g., a pipeline processing adder tree or group of parallel counters, allowing the simultaneous summation of many ray sum contributions, each pixel would be generated within a single operating cycle of the processor; in addition, no read operations and only n^2 write operations into the image memory are required. (See Gilbert et al., 1979, and Swartzlander et al., 1978). Hence, with reference to fixed memory cycle times, back projection executed as a parallel computational process is potentially (2M-1) times faster than its sequential counterpart. Because the parallel processing back-projection unit may be fabricated from very few hardware subunits of different design, each replicated a number of times, there is far less than a (2M-1) times increase in cost for the additional engineering design effort to achieve this increase in throughput. Because of its inability to perform the same operations on multiple operands simultaneously, a general purpose computer can neither mechanize nor exploit the computational parallelism inherent in the pixel-driven back-projection algorithm.

3. FLEXIBILITY OF THE HIGH SPEED PARALLEL PROCESSING RECONSTRUCTION ARCHITECTURE

In view of the reliance of the design of Figure 2 upon reprogrammable memory to initiate and control all phases of algorithm execution, the design is referred to as a "memory-driven" processor. This mechanization, in combination with the pixel-driven concept of back projection, possesses a substantial degree of operational flexibility resulting from the ability of the user to redefine the step-by-step operation of the processor by modifying the contents of the projection and projection address memories; this implementation also assures that alterations in the intended function of the processor will require neither rewiring nor compromise its throughput. Several of the applications of such a design are described here.

The architecture of Figure 2 is not constrained to back-projection of the identical number of projections as the number of projection memories implemented in hardware. Back-projection of fewer projections than the number of projection memories requires only that the contents of the projection address memory be loaded with a special "null address" which in effect suppresses the contributions from those projection memories not to be used during a back-projection operation. Conversely, a greater number of projections than the number of projection memories may be back-projected in multiples of the number of projection memories. The first set of projections is processed in the manner described earlier; subsequent sets of projections are back-projected pixel-by-pixel into the same image space by employing a "read-add-write" method similar to that described for projection-driven back-projection. The read-add-write technique can be repeated for as many sets of projections as desired, provided only that the operand resolution of the image memory is sufficiently large to preclude overflow or truncation error during the execution of successive summations into each memory location.

The pixel-driven special purpose processor is capable of generating reconstructions from projections formed either with parallel or with fan shaped beams of x-rays simply by reloading the contents of the projection address memories. Regardless of whether the projection data are collected with a parallel or a fan beam of x-rays, it is only necessary to identify those rays which intersect a given pixel undergoing back-projection, thereby establishing the correct indices of those ray sums whose contributions will be back-projected into the pixel. In the hardware implementation described here, the index values are represented as a set of memory addresses which are used to read the appropriate ray sum values from the projection memories. Thus, only the contents of the projection address memory need be altered when converting from fan beam to parallel-beam back-projection; for the parallel ray case, the multiplication of the ray sums by the "back-projection weights" must also be suspended.

It is frequently desirable to reconstruct a subarea of the entire cross section of the structure under study, either to exclude objects outside the region of interest, or more usually, to achieve increased spatial resolution within a subregion of the cross section. The table-driven architecture of Figure 2 readily accommodates the execution of such "zoom" back-projections, particularly when each filtered projection contains a large number of ray sums (perhaps provided via interpolation). The projection address memory can be loaded with the appropriate ray sum indices to back-project a higher resolution image representing any preselected portion of the entire cross section. Any desired degree of magnification can be achieved in an arbitrary subregion of the entire cross section up to the resolution limit established by the information content and signal/noise ratio of the projections, provided that the total number of back-projected pixels does not exceed the number of locations in the image or projection address memories.

As originally proposed by Herman et al. (1976), the fan beam convolution algorithm requires a set of projections equispaced about 360° of the object to be reconstructed. Herman and Naparstek (1977) have also suggested a modification of the fan beam algorithm to permit reconstructions from projections equispaced over an angular range of 180° plus the divergence angle α of the x-ray beam. This form of the algorithm is well suited to the source-detector geometry and rotational constraints of the DSR gantry which, because of its low rate of angular rotation, would largely sacrifice temporal resolution if required to record data from the entire circumference of the object. The Herman-Naparstek modification of the fan beam convolution algorithm requires that selected ray sums from projections recorded near 0° and $(180 + \alpha)$° angular displacements of the x-ray source not be back-projected at all, i.e., are assumed to have zero magnitude. Since the indices of these zero-magnitude ray sums may be calculated in advance, the pixel-driven back-projector can directly exploit this modified algorithm by defining as identically zero the contents of one address location in each projection memory; the projection address memory is loaded with the addresses of these null locations for ray sums to be ignored during back projection, thereby directly satisfying the Herman-Naperstek modification.

If the data base of x-ray projections represents a three-dimensional volume of tissue, the pixel-driven reconstruction method, in combination with the special purpose reconstruction processor architecture of Figure 2, possesses several useful operational flexibilities in addition to the increased throughput capabilities expected from a parallel computational process. The projection address memory of Figure 2 allows the pixel-driven back-projector to reconstruct directly any desired coronal, sagittal, or oblique section (or even nonplanar sections) through the three-dimensional image space which may be generated from the projection data. Conceptualizing the three-dimensional image as a stack of parallel transverse cross sections, the intersection of an oblique cutting plane with any of the transverse sections is a straight line segment in the plane of the transverse section. Accounting for the finite pixel dimensions of the image volume, each region of intersection between an oblique and a transverse section, both of finite thickness, is a narrow stripe approximately one pixel in width within the planar slab of the transverse section. Presently, if the reconstruction of an oblique section is desired, projection-driven algorithms prepared for general purpose computers back project all transverse sections within the image volume, and then incorporate the appropriate pixels from each transverse section into a two-dimensional representation of the oblique section. However, since only a few pixels from each transverse section are eventually included in the oblique image, it would be sufficient to back-project only those pixels which form a portion of the oblique section. The table-driven back-projection hardware of Figure 2 is well suited to direct reconstruction of non-transverse planes, since the projection address memory need only be loaded with the ray sum indices necessary to reconstruct the appropriate pixels from each transverse section; a minimal amount of additional memory-based microprogram control hardware would allow the processor to be loaded initially with the ray sum indices necessary to reconstruct an oblique section directly and, with a sufficiently detailed sequence of microcontrol instructions, to specify the procedure completely. The reconstruction processor could then automatically cycle entirely through the oblique view reconstruction without further external intervention.

The design proposed here for a parallel processing special purpose computer for three-dimensional x-ray tomographic reconstruction of data from the Dynamic Spatial Reconstructor must be evaluated for operational flexibility and ease of application in a realistic computational environment. Accordingly, a prototype parallel processing filtered back-projection reconstruction processor with modular replaceable subsections has been designed and fabricated, and is currently undergoing operational verification (Figure 4). The modularly of this prototype, which implements the architecture of Figure 2, will permit direct comparisons of a variety of designs for the most critical subassemblies of the parallel processing computer.

Fig. 4. Engineering prototype special purpose cross-sectional
reconstruction processor fabricated with commercially available
components. Leftmost logic cards are direct convolution filter.
Rightmost cards are four projection memory subunits, projection
address and image memories, and system control. Design is that
of Figure 2. Note the pair of very large scale integrated circuit
16x16 multiplier components on extended logic card.

4. ACKNOWLEDGEMENTS

The technical assistance of the following individuals at the
Mayo Clinic is acknowledged: R. D. Beistad, D. A. Brumwell,
A. Chu, and M. T. Despres; for assistance in the preparation of
text and figures: D. C. Darling, E. C. Quarve, M. C. Fynbo,
P. A. Monson, M. A. Engesser, E. A. Gjellstad, I. E. Donovan.
Contributions to this work by E. E. Swartzlander, Jr. of the TRW
Defense and Space Systems Group (Redondo Beach, California, USA)
have also been invaluable.

This investigation was supported in part by United States
Public Health Service Granst HL-04664 and RR-00007 from the
National Institutes of Health, and grants from the Fannie E.
Rippel Foundation and Control Data Corporation.

5. REFERENCES

Bracewell, R., The Fourier Transform and Its Applications, New York,
    McGraw Hill Book Company (1965), Ch. 2.

Brooks, R. A., and DiChiro, G., "Principles of Computer Assisted Tomography (CAT) in Radiographic and Radioisotopic Imaging," Phys. Med. Bio. 21:689-732 (1976).

Gilbert, B. K., Chu, A., Atkins, D. E., Swartzlander, E. E., Jr., and Ritman, E. L., "Ultra High-Speed Transaxial Image Reconstruction of the Heart, Lungs, and Circulation via Numerical Approximation Methods and Optimized Processor Architecture," Computers and Biomed. Res. 12(1):17-38 (1979).

Gilbert, B. K., Storma, M. T., Ballard, K. C., Hobrock, L. W., James, C. E., and Wood, E. H., "A Programmable Dynamic Memory Allocation System for Input/Output of Digital Data into Standard Computer Memories at 40 Megasamples/Second," IEEE Trans. Comput. C-25(11):1101-1109 (1976).

Groginsky, H. L., and Works, G. A., "A Pipeline Fast Fourier Transform," IEEE Trans. Comput. C-19(11):1015-1019 (1970).

Herman, G. T., Lakshminarayanan, A. V., Naperstek, A., Ritman, E. L., Robb, R. A., and Wood, E. H., "Rapid Computerized Tomography," in Medical Data Processing (Laudet, M., Anderson, J., and Begon, S., eds.), London, Taylor and Francis (1976), pp. 581-589.

Herman, G. T., and Naparstek, A., "Fast Image Reconstruction Based on a Radon Inversion Formula Appropriate for Rapidly Collected Data," SIAM J. Appl. Math. 33(3):511-533 (1977).

Kenue, S. K., and Greenleaf, J. F., "High Speed Convolving Kernels Having Triangular Spectra and/or Binary Values," IEEE Trans. Nuc. Sci. NS-26(2):2693-2696 (1979).

Ramachandran, G. N., and Lakshminarayanan, A. V., "Three Dimensional Reconstruction from Radiographs and Electron Micrographs; Application of Convolutions Instead of Fourier Transforms," Proc. U.S. Natl. Acad. Sci. 68:2236-2240 (1971).

Reed, I. S., Kwoh, Y. S., Truong, T. K., and Hall, E. L., "X-Ray Reconstruction by Finite Field Transforms," IEEE Trans. Nuc. Sci. NS-24(1):843-849 (1977).

Ritman, E. L., Robb, R. A., Johnson, S. A., Chevalier, P. A., Gilbert, B. K., Greenleaf, J. F., Strum, R. E., and Wood, E. H., "Quantitative Imaging of the Structure and Function of the Heart, Lungs, and Circulation," Mayo Clinic Proceedings 53:3-11 (1978).

Robb, R. A., Ritman, E. L., Harris, L. D., and Wood, E. H.,
    "Dynamic Three-Dimensional X-Ray Computed Tomography of the
    Heart, Lungs, and Circulation," IEEE Trans. Nuc. Sci. NS-26
    (1):1646-1660 (1979).

Swartzlander, E. E., Jr., Gilbert, B. K., and Reed, I. S., "Inner
    Product Computers," IEEE Trans. Comp. C-27(1):21-31 (1978).

COMPUTER ANALYSIS OF THE ULTRASONIC ECHOCARDIOGRAM

M. Kuwahara, S. Eiho, H. Kitagawa, K. Minato and N. Miki

Automation Research Laboratory

Kyoto University, Uji, JAPAN

1. INTRODUCTION

Ultrasonic echocardiography (UCG) has been widely used as a noninvasive, safe and reliable diagnostic tool in clinical medicine. Measurement techniques using ultrasound have been proposed during the past decade. In using a UCG, a short burst of ultrasound is emitted through the body surface from a transducer. Echoes reflected at the borders between two media of different acoustic impedance are caught using the same transducer as a receiver. An M-mode display of the ultrasonic echocardiogram gives much useful information on the internal structure and dynamic characteristics of the heart. However, ultrasonic echocardiograms are usually stored as data in a strip chart or as photography and the analysis of the data to evaluate the cardiac function quantitatively is performed manually by medical doctors consuming much of their time and energy.

There are published works on direct transfer of UCG signals into a digital computer. For example, Romic and Hagan (1974) used a special purpose tape recorder which records UCG signals at high speed and plays back the signals at a low speed for computer processing. Hirsh et al. (1973) also used a special transient recorder to record UCG signals. In both papers, the UCG signals are digitized as 4-bit words in the last stage of signal analysis and the processing of UCG signals are automatically performed using the computer. But there are objections to their methods: (1) The need for special equipment to collect the data and (2) the inexactness of the digitized data.

In this paper, we deal with a system and a method by which the UCG signals are directly transferred from a conventional UCG instrument to a minicomputer in real-time and stored at the same time on a video tape. The analog UCG signals from the UCG instrument or the video tape are converted into 8-bit digital values for automatic processing by the minicomputer. Cardiac structures are delineated by tracing the UCG signals from the interventricular septum and the posterior wall of the left ventricle. Then various numerical values are calculated giving a quantitative description of cardiac function.

2.   SYSTEM CONFIGURATION

Our system uses a minicomputer (HP2100A, 32Kw) with a disc operating system, a graphic CRT display (Tektronix 4012), a conventional UCG instrument, an AD converter, shift registers (used as buffer memories), a video tape recorder (VTR) and special interface circuits which connect the UCG instrument to the minicomputer and the video tape recorder.

The UCG signals from the UCG instrument are digitized in real time through the interface circuit and transferred to the minicomputer. The UCG signals are also stored as electric signals on the video tape. Thus we can observe the UCG signals repeatedly on the oscilloscope of the UCG instrument and transfer the signals to the minicomputer, digitizing them in the same way as in the direct real-time transfer.

Figure 1 is a block diagram representing the system. The circuits enclosed in dotted lines are for storage transfer of the UCG signals from the video tape to the minicomputer. Therefore, the circuits need not be used when the UCG signals are transferred directly from the UCG instrument to the minicomputer.

The transducer of the UCG instrument in our case emits a pulse train of ultrasound using a frequency of 2.25 MHz and pulses about 2μs in duration at intervals of 0.5ms. During each interval, echoes are caught by the same transducer. These echoes have a bandwidth of 1MHz and a dynamic range of about 50dB. After a low-pass filtering using a 1MHz cut-off frequency, we use an AD converter that has a sampling rate of 2MHz and an 8-bit resolution. This converter digitizes every 4th echo into 512 8-bit words in the usual case. We set the selection rate for the echoes from each 4th, 8th, 16th, 32nd or 64th echo by means of a switch. The digitized data thus obtained are transferred to shift register and then transferred from the shift registers to the minicomputer packed into 16-bit words at a computer compatible rate less than the sampling rate. These transfers must be finished before the emission of the next pulse.

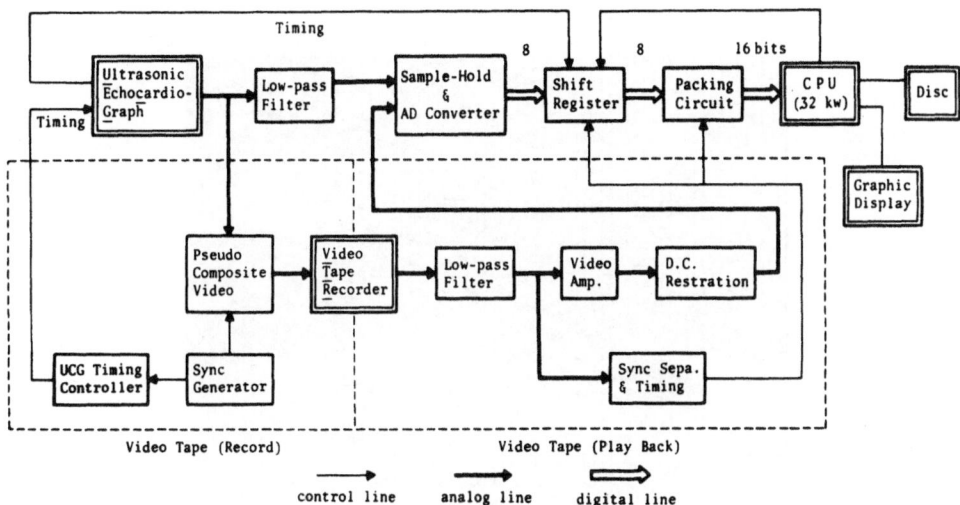

Fig. 1. Block diagram of the system.

     To record the UCG signals on a video tape, a conventional UCG
instrument must be slightly modified so that the burst of ultrasound is synchronized with the external timing pulse emitted by
every 8th horizontal TV sync pulse (8th sync). The TV sync pulse
are added to the UCG signals and we can thus obtain composite
video signals which are recordable by the VTR. The transfer to
the minicomputer of the UCG signals recorded on video tape in this
manner is performed in the same way as in the direct real-time
transfer of echo signals. On the other hand, the UCG signals recorded on the video tape can be displayed on the oscilloscope of
the UCG instrument for repetitive observation by medical doctors.

3.   PROCESSING OF ULTRASONIC ECHOCARDIOGRAM

     If we can clarify the internal structure of the heart through
the digitized UCG signals of the M-mode display, it is easy to
obtain various numerical values relating to the left ventricle,
e.g., the length of the short axis (diameter), the thickness of
the posterior wall, the velocity of circumferential fiber shortening, and the ventricular volume. These values are very important
in evaluating cardiac functions. Therefore, we must start by detecting the edges of the interventricular septum and the posterior
wall of the left ventricle from the UCG signals.

     Figure 2 shows an example of an original digitized M-mode UCG
signal recording, which is stored on the disc of the minicomputer
and displayed on the CRT. In this figure, we can see the right
ventricle, the interventricular septum, the left ventricle, both
the endocardium and the epicardium of the posterior wall, and the
lung tissue. The curve on the right side of this figure shows the

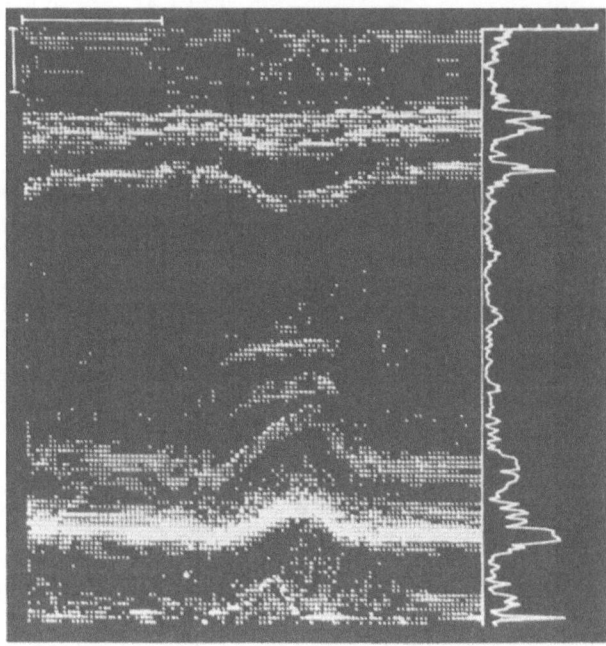

Fig. 2. An example of the digitized M-mode ultrasonic echocardiogram.

A-mode display of the last line of the M-mode display. The M-mode
display of the UCG signals shows the relation between the depth of
each tissue from the chest wall as a function of elapsed time. On
the other hand, the A-mode display shows more clearly the relative
intensities of the echo signals. In the A-mode display we may
observe the interventricular septum, endocardium, and epicardium.
However, edge lines are not always so clear and special techniques
to detect edge lines may be required.

    The waveforms shown in the UCG M-mode display indicate the
movements of cardiac structures. The instantaneous position of
each structure is measurable by detecting the maximum intensity of
echoes in the A-mode display in the appropriate area. It can be
assumed that the position of the structure is within a narrow
region for successive vertical lines of the M-mode UCG signal,
because the movement of each structure in a short time is limited.
Let us take a vertical line in an M-mode echocardiogram and assume
the position of the structure on the line to be d_{i-1} as shown in
Figure 3. The position of the structure on the next vertical line,
d_i, will be decided within a limited region between $d_{i-1} - w_n$ and
$d_{i-1} + w_p$, where w_n and w_p are selected in consideration of the

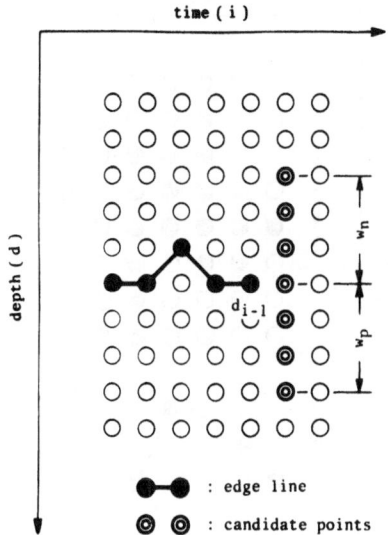

Fig. 3. Limited region of search for detecting cardiac structures.

velocity of the movement and the displacement of the structure and
other factors which have a bearing on the movement of the structure.
Thus, the maximum value of echo intensities in the selected region
is taken as the position of the structure.

In general, the epicardium of the posterior wall of the left
ventricle is clearly detectable and the echo from it frequently
has the highest value among echoes from all tissue boundaries.
On the contrary, the echo from the endocardium of the left ventricle
is relatively indistinct and, at times, broken at places. Thus the
endocardium should be located by tracing the movement of the epicardium, because the endocardium and epicardium are closely related
in front of and behind the posterior wall of the left ventricle.

The region of search for detecting the position of the endocardium does not have a constant width. It is broadened in the
direction of movement of the epicardium which has already been
detected. The amount of broadening is selected to be proportional
to the velocity of movement of the epicardium as shown in Figure 4.

Similarly, the interventricular septum of the left ventricular
side is clear and easily detectable. The septum of the right ventricular side is indistinct and difficult to detect, as in the case
of the endocardium, except by tracing the movement of the septum
of left ventricular side. The starting point from which to search
for each structure can be automatically fixed by summing up the
echoes of several vertical lines in the M-mode echocardiogram.

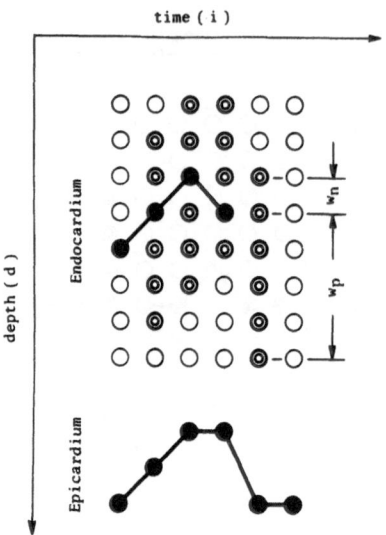

Fig. 4. Region of search for detecting the endocardium.

4.   CLINICAL RESULTS

Figure 5 shows the cardiac structures detected by fixing the
values of w_n and w_p, which limit the region of search, at 4 points.
The upper two lines show the edges of the interventricular septum
and the lower two show the movement of endocardium and epicardium
of the left ventricular posterior wall. Figure 6 shows the detected
cardiac structures superimposed on the original digitized UCG M-mode
signals. From these detected lines we can obtain various information on cardiac function as shown in Figures 7-10. Figure 7 gives
the velocity of movement of the epicardium. In Figure 8 changes of
the diameter and the posterior wall thickness of the left ventricle
are given. Figure 9 shows the velocity change of circumferential
fiber shortening of the left ventricle and Figure 10 gives the
volume change of the left ventricle calculated by Bigson's formula
with the change of diameter. These clinical results are for a person with a healthy heart. This information can be effectively combined in the color-coded displays shown in Color Plates 17 and 18.*

Figure 11 is an example for a patient with an atrium septal
defect (ASD). The detected structures are superimposed on the
M-mode display of the UCG signals. We can see clearly the para-
doxical movement of the interventricular septum in systole, which
is considered as a characteristic of ASD.

Finally, let us compare the quality of UCG signals digitized
directly from the UCG instrument with those digitized through the

*The color plates will be found following page 144.

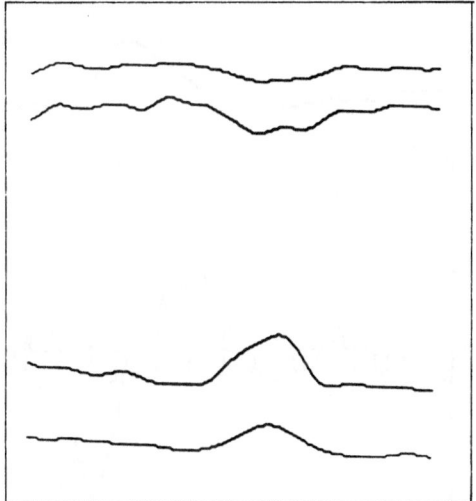

Fig. 5.   Cardiac structures detected in the M-mode echocardiogram.

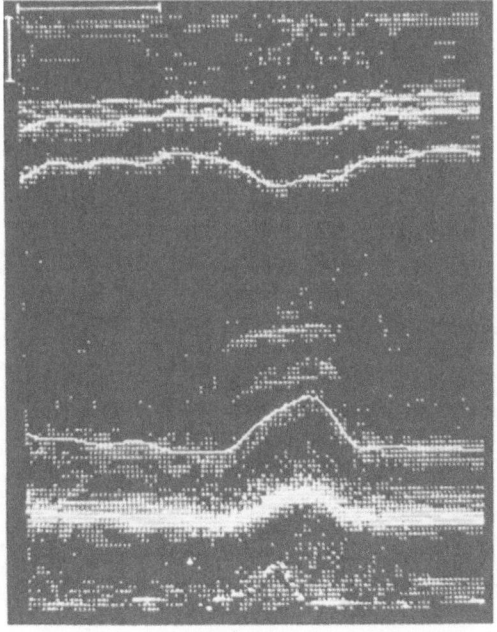

Fig. 6.   Detected cardiac structures superimposed on the original
ultrasonic echocardiogram.

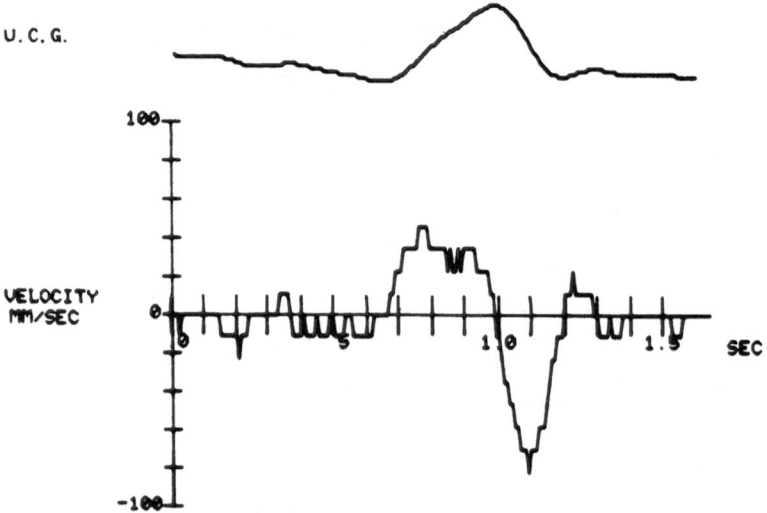

Fig. 7. Velocity of the movement of endocardium.

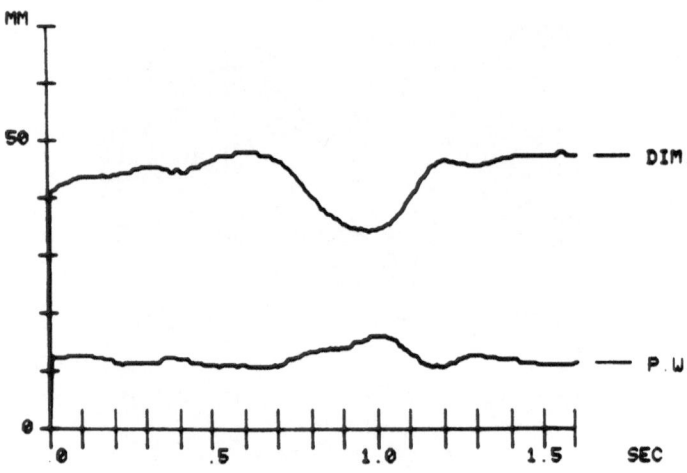

Fig. 8. Changes in the diameter (DIM) and the posterior wall
thickness (P.W.) of the left ventricle.

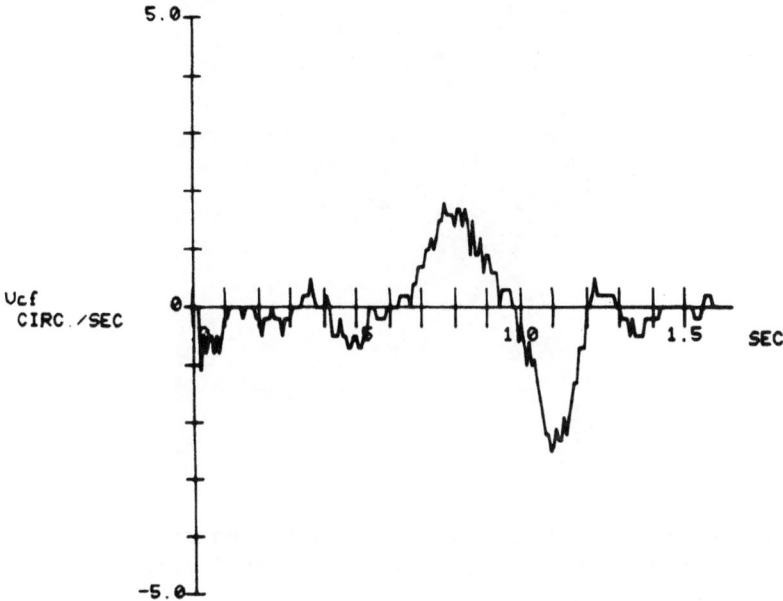

Fig. 9. Velocity change of the circumferential fiber shortening
of the left ventricle.

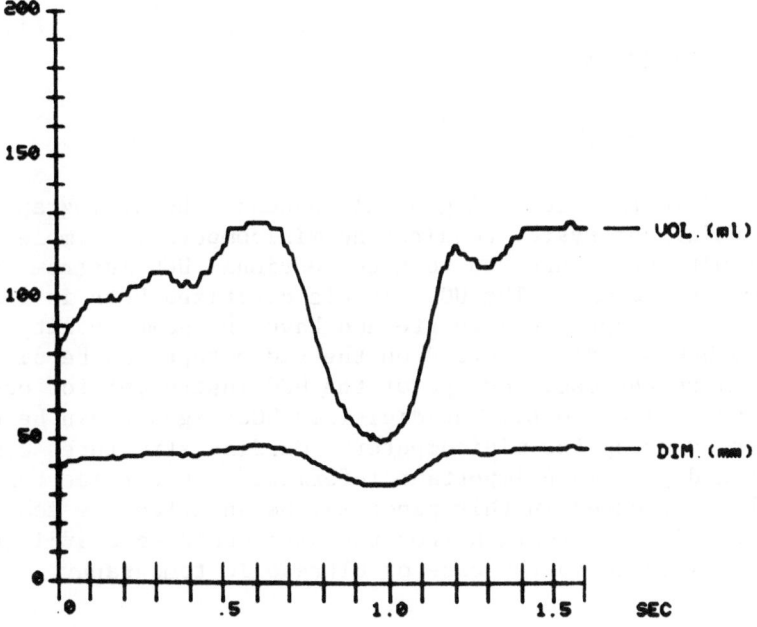

Fig. 10. Volume change (VOL) of the left ventricle calculated
using Gibson's formula and the change of diameter (DIM).

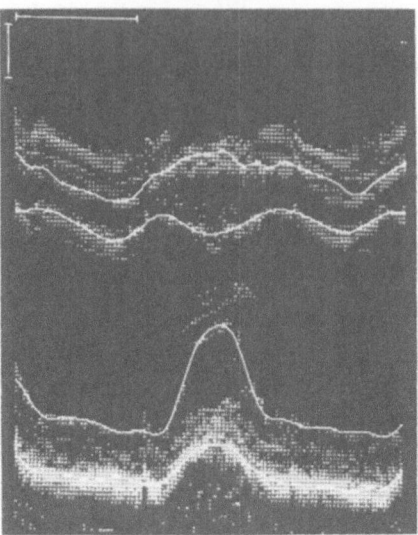

Fig. 11. An M-mode display with the detected cardiac structures
indicated for a patient with an atrium septal defect.

VTR and interface circuits. Figure 12a shows an M-mode display of
UCG signals on a CRT obtained by real time direct digitization of
the movement of mitral valve for a normal subject. Figure 12b
gives the similar display obtained through the video tape for the
same subject. These two figures can be seen to be very similar and
to have equal quality.

5.   CONCLUSIONS

     The video tape recording of ultrasound echocardiographic (UCG)
signals using our system requires no minicomputer. Simple interface circuits are connected to a conventional UCG instrument to
make such recordings. The UCG signals digitized both in real-time
and from video tape are accurate and have the same quality. Moreover, the UCG signals recorded on the video tape can be displayed
repeatedly on the oscilloscope of the UCG instrument for observation by medical doctors. The digitized UCG signals can be easily
processed by using the minicomputer. Cardiac structures can be
detected and give much important information on cardiac function.
The method presented in this paper may be an effective tool for
getting cardiac information from the ultrasonic echocardiogram and
also be applicable to the case of ultrasonic tomography.

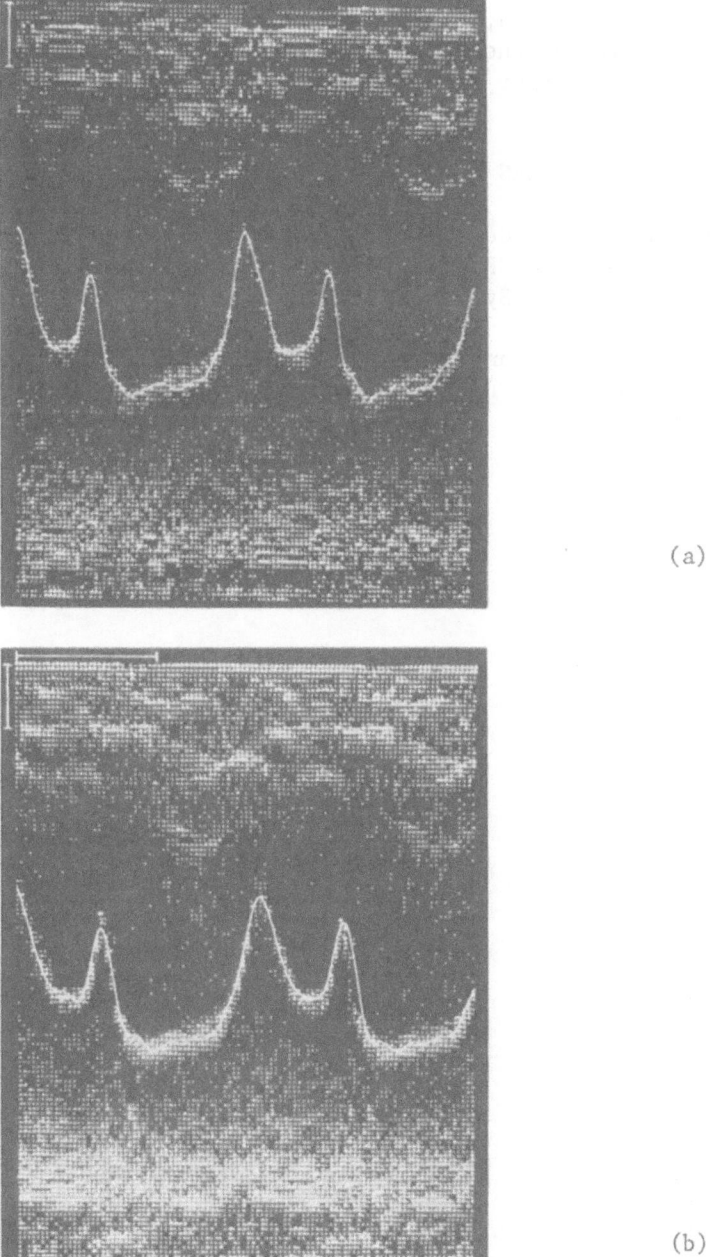

(a)

(b)

Fig. 12. Comparison of the quality of the directly digitized
ultrasonic echocardiogram (a) with that obtained through video
tape recording (b).

6.   ACKNOWLEDGEMENTS

The authors thank Profs. C. Kawai and A. Hirakawa, and
Drs. H. Kotoura, S. Sasayama and G. Osakada of Kyoto University
Hospital for their cooperation. This research was supported in
part by a grant from the Ministry of Health and Welfare of Japan.

7.   REFERENCES

Hirsh, M., Sanders, W. J., Popp, R. L., and Harrison, D. C.,
    "Computer Processing of Ultrasonic Data from the Cardio-
    vascular System," Comput. Biomed. Res. 6:336 (1973).

Romic, C. A., and Hagan, A. D., "Automated Echocardiogram Analysis,"
    Proc. San Diego Biomedical Symposium (1974), p. 145.

TOWARD COMPUTED DETECTION OF NODULES IN CHEST RADIOGRAPHS

J. Sklansky, P. V. Sankar, M. Katz, F. Towtig,
D. Hassner, A. Cohen and W. E. Root

University of California

Irvine, CA 92717, USA

1. INTRODUCTION

The detection of pulmonary nodules is recognized by radiologests as an important segment of the analysis and diagnosis of chest radiographs. Garland (1950) and others have demonstrated that radiologists routinely miss about 30% of the abnormalities visible in chest film radiographs. Ziskin et al. (1971) show that radiologists produce about as many or more false positives than false negatives when they search for a solitary nodule in a chest radiograph.

We are developing an interactive computer system with the objective of reducing the rate by which radiologists miss lung tumors to below 5% as well as reducing the number of false positives. The system consists of a scanning microdensitometer, a digital minicomputer, a computer program, and a digital cathode-ray-tube color display. This system detects nodules in the range 5mm to 25mm and classifies them into two classes: *nodule* and *nonnodule*.

Current results are quite promising: in a set of 5 radiographs containing 14 nodules, a few of which are difficult for competent radiologists to detect, our computer system misses only one and generates an average of seven false positives per radiograph. Almost all of these false positives can be correctly screened by a competent radiologist.

2.   MATERIALS AND METHODS

Each radiograph is scanned on an Optronics rotating-drum micro-densitometer at an aperture width of 200µm and an interpixel spacing of 200 µm, under control of a Hewlett Packard model 2114B minicomputer. This yields a *finely sampled* digitized image consisting of a 1536x1536 array of 8-bit pixels. Subsequent analysis takes place on a Perkin-Elmer model 7/32 minicomputer, with 320 kilobytes of main memory. To save computation time almost all of this analysis is carred out in integer arithmetic (i.e., without floating point arithmetic).

A *coarsely sampled* digitized image is obtained from the finely sampled image by replacing every 6x6 square subarray of pixels by the average of the gray values in that array. The coarsely sampled image is a 256x256 array of 8-bit pixels. This image is "pre-processed" in four steps: (1) normalized zonal notch filter, (2) gradient, (3) threshold, and (4) straight-line suppression. The preprocessed, coarsely sampled image enters a circle detector. This circle detector finds objects, within a prescribed range of sizes, whose boundaries are nearly circular or contain nearly circular arcs. The output of the circle detector enters our heuristic boundary follower. While computing the boundary, descriptors of the shape of the boundary are also computed. These descriptors are entered as "features" in a simple classifier. The output of this classifier has one of two values: *nodule* and *nonnodule.*

A Genisco graphics display system produces a full size enhanced black-and-white image of the chest radiograph in a 512x640 array of dots on a 25-inch Conrac monitor. Colored plus signs ("+") mark the computed candidate nodules.

2.1 Preprocessor

Our preprocessor is designed to carry out the following functions: (a) equalize the wide range of intensities or edges of objects in the images, (b) detect local edge elements of "strokes," (c) suppress false edge elements caused by noise, and (d) suppress short straight edge sections that are likely to belong to non-nodules.

Edge equalization is carred out by our normalized zonal notch image filter. This filter selects a "zone" of pixels whose gray levels are near that of the center pixel within a 10mm-wide window. The filter subtracts the local average of this zone, and divides the result by the standard deviation of the gray levels in the zone. Thus, this filter is a combination of the zonal notch filter of Schwartz and Soha (1977) and the statistical differencer described

in Pratt (1978). The use of the zone in this filtering process
tends to preserve edges between regions of widely differing average
gray levels. Edge detection is carried out by a Sobel gradient
operator (see Pratt, 1978), yielding the digital vector field
$g(x)$.

False edge elements are suppressed and further edge equalization is carried out by the following "adaptive thresholding" of the
modulus of $g(x)$. (We denote this modulus by g(x).)

$$g_a(x) = \begin{cases} g(x) - k\ \overline{g}(x) & \text{if } g(x) > k\ \overline{g}(x) \\ 0 & \text{otherwise} \end{cases}$$

where $x = (x_1, x_2)^T$ = position vector in the image space, and $\overline{g}(x)$
is the mean of $g(x)$ in a 9x9 neighborhood. Currently we use k = 1.

Straight line segments in $g_a(x)$ are suppressed by first detecting straight lines in a set of overlapping 20x20 pixel windows.
These straight lines are detected by peaks in the projections of
the elements of $g_a(x)$ in these windows along a set of orientations
separated by seven-degree intervals. [This technique is similar to
Griffith's (1973).] A weighted subtraction of the detected lines
from $g_a(x)$, followed by a thresholding for noise suppression, yields
the line suppressed gradient, $g_p(x)$. This is the output of the preprocessor.

2.2 Circularity Detector

Our circularity detector finds digital arcs that are circular
or nearly circular in the line-suppressed gradient, $g_p(x)$. This
detector is our extension of the Hough line detector to the detection of circular arcs as presented in Kimme et al. (1975). Our
method improves on Hough's concept by exploiting the directional
information in $g_p(x)$. The output of the circularity detector is a
set of circle centers and associated radii: $\{(c_i, R_i)\}$.

2.3 Boundary Follower

The circle centers and radii produced by the circularity
detector are sent to the heuristic boundary follower. Here each
pair (c_i, R_i) generates an annular-ring-shaped potential well in
the finely sampled image. The potential well is used as a plan
for guiding a heuristic search in the manner of Ballard and
Sklansky (1976). The search space is partitioned into four quadrants, centered at c_i. The boundary follower consists of a sequence

of heuristic searches, each search confined to a single quadrant.
Breaking the search space into quadrants provides a strategy for
filling very large gaps in the boundary--as great as 100 degrees
or more, and it reduces the time for search.

In each quadrant the candidates for the boundary are grown as
a tree in which the pixel at each node, except the root, has a
unique predecessor corresponding to the current minimal-cost path
through that pixel. This ensures that the number of possible paths
is bounded by the number of pixels in the quadrant. (Contrast this
with the exponentially increasing number of candidate paths in
earlier heuristic boundary followers of Montanari (1971), Martelli
(1976), and of Ashkar and Modestino (1978). Our cost function at
point x on a path depends on the accumulated moduli of $g_p(x)$ along
the path, the agreement of the direction of $g_p(x)$ with the plan's
direction, the value of the plan's potential well at position x,
and the distance of x from the second radial edge of the current
quadrant. For efficiency, our tree search is implemented in a
heap data structure as described in Knuth (1973).

In the first quadrant the starting points of the heuristic
search are chosen along the positive portion of the vertical axis
(the x_2 axis). The ingredients entering into this choice are
$g_p(x)$ and the position of the candidate start point in the plan.
Using our cost function we grow paths counterclockwise in the first
quadrant. The end points of these paths lie on the second radial
edge of the first quadrant, namely the negative portion of the
horizontal axis. These end points are the start points for the
heuristic search in the next quadrant. This process is repeated
five times, once for each quadrant. The "fifth" quadrant is a
second search in the first quadrant, using the end points of the
fourth quadrant as start points for the second search of the first
quadrant. The second search in the first quadrant is needed,
because the start points in the first search of the first quadrant
were chosen arbitrarily.

This technique incorporates what we believe are the best
features of the earlier versions of heuristic search boundary followers. We find that it is computationally more economical than
those boundary followers, as a result of its use of an expected
cost function and the heap data structure for implementing the
minimum cost tree search. Figure 1 illustrates the performance of
the boundary follower by showing the computed boundaries for a set
of two nodules and two false positives (nonnodules).

2.4 Classifier

After examining the computed boundaries of the false and true
positives, such as those in Figure 1, we chose the following three

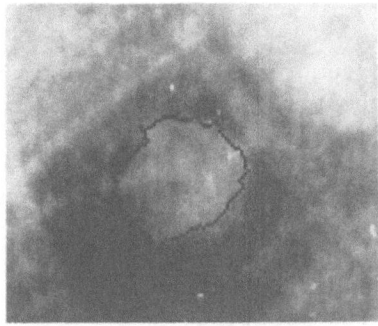

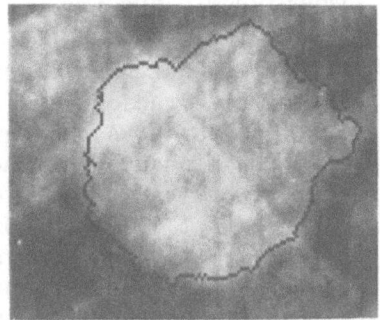

True positive   True positive

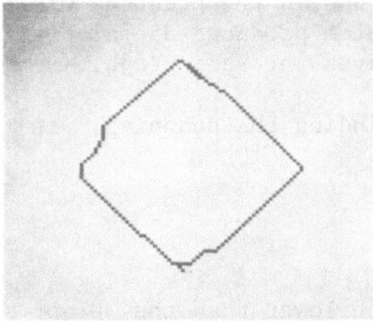

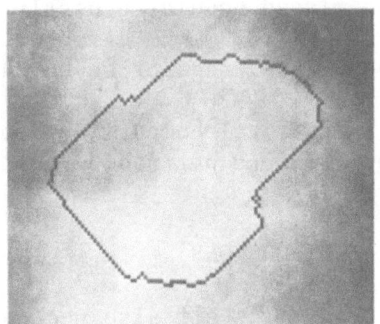

False positive   False positive

Fig. 1. Examples of boundaries of true and false positives found by the circle detector.

shape parameters as features for a classifier. This classifier labels each computed boundary as *tumor* or *nontumor*.

    1. Number and length of straight segments of the boundary.

    2. Search time.

    3. Cost of the boundary.

These features were chosen not only for their discriminatory power, but also because they are obtained almost as a byproduct of our boundary follower.

    The two decision regions of the classifier are an octant and its complement, the octant corresponding to *tumor*. The octant is formed by three planes, each of which is perpendicular to one of the coordinate axes.

3.   EXPERIMENTAL RESULTS

We applied our nodule detector and tumor-nontumor classifier
to a set of five radiographs: two PA chest radiographs of human
subjects, and three PA radiographs of Dr. J. G. Pearce's anthropomorphic chest phantom at the UCI Department of Radiological
Sciences. These radiographs contained 14 nodules, with diameters
in the range from 4.9mm to 20mm. The modules in the chest radiographs were verified as tumors by A. Cohen and W. Root of UCI.
The nodules in the chest phantom were simulated tumors implanted
in the phantom at specified and documented positions.

Among these 14 nodules, our system detected 13 and missed one.
(The missed nodule, a nearly invisible one in the phantom, was the
smallest: 4.9mm in diameter.) Our system produced 35 false posi-
tives--an average of seven false positives per radiograph.

The total computing time, not including the scanning, was
about 2 hours per radiograph.

4.   CONCLUSIONS

Our rate of false negatives is much lower than the 30% or so
typically produced by practicing radiologists. The rate of false
positives may be tolerable, since the radiologist who uses our
system can screen candidate nodules visually with little difficulty.

Nevertheless, we are working to reduce both the false negative
rate and the false positive rate. Our main goal is to reduce to
5% or less the rate of false negatives for tumors of 5mm in diameter
or larger.

Clearly in a system in which the algorithms are carried out in
"real-time" the computing time must be reduced by two orders of
magnitude, i.e., by a factor of 100. By present technology we can
reduce this time by a factor of 10, using specialized software and
faster hardware. We believe the remaining factor of ten will be
achievable by advances in digital computer hardware and by specialized hardware implementations of parts of our software.

5.   ACKNOWLEDGEMENTS

We are grateful to J. G. Pearce and G. D. Gillan for designing
and fabricating the simulated nodules, implanting them in the
anthropomorphic chest phantom, and providing us with documented
radiographs of various distributions of these simulated nodules in
the chest phantom, and to A. Cohen and W. Root for visually analyzing
the radiographs in our data base.

We are indebted to D. Janney and J. R. Breedlove of Los Alamos Scientific Laboratory, Los Alamos, New Mexico, for providing us with a high-quality digitization of one of the chest radiographs in our data base.

This research was supported by the National Institute of General Medical Sciences of the U. S. Public Health Service under Grant No. GM-17632 and by the National Science Foundation under Grant No. ENG77-17081.

6. REFERENCES

Ashkar, A. P., and Modestino, J. W., "The Contour Extraction Problem with Biomedical Application," Comput. Graphics and Image Proc. 7:331-355 (1978).

Ballard, D. H., and Sklansky, J., "A Ladder Structured Decision Tree for Recognizing Tumors in Chest Radiographs," IEEE Trans. on Computers C-25(5):503-515 (1976).

Garland, L. H., "On the Reliability of Roentgen Survey Procedures," Amer. J. Roent. 64(32) (1950).

Griffith, A. K., "Edge Detection in Simple Scenes Using *a priori* Information," IEEE Trans. on Computers C-22(4):371-381 (1973)

Kimme, C., Ballard, D. H., and Sklansky, J., "Finding Circles by an Array of Accumulators," Comm. of the ACM 18(2):120-122 (1975).

Knuth, D. E., The Art of Computer Programming--Vol. 3/Sorting and Searching, Addison-Wesley Publishing Co. (1973).

Martelli, A., "An Application of Heuristic Search Methods to Edge and Contour Detection," Comm. ACM 19:73-83 (1976).

Montanari, V., "On the Optimal Detection of Curves in Noisy Pictures," Comm. ACM 14:335-345 (1971).

Pratt, W. K., Digital Image Processing, New York, Wiley and Sons, (1978).

Schwartz, A. A., and Soha, J. M., "Variable Threshold Zonal Filtering," Applied Optics 16(7):1779-1781 (1977).

Ziskin, M. C., Shea, F. J., Kundel, H. L., and Revesz, G., "Accuracy of Radiologists' Decision Making," in Quantitative Imagery in the Biomedical Sciences, Photo-optical Instrumentation Engineers, May 1971.

A PARALLEL PROCESSING SYSTEM FOR THE THREE-DIMENSIONAL DISPLAY OF
SERIAL TOMOGRAMS

K. Tanaka and S. Tamura

Faculty of Engineering Science, Osaka University

Osaka, 560, JAPAN

1. INTRODUCTION

The soft tissue visualization capability of ultrasound has been
used effectively for clinical diagnoses throughout the world. By
using an ultrasonic imaging system, a two-dimensional image of a
section of the human body is obtained with little or no known
hazard to the patient.

An ultrasonic beam scans an entire plane and recorded pulse-echo
signals produce a picture whose intensity is proportional to the
reflectance of the tissue at corresponding points in the image
plane. Thus a tomogram or a B-mode image is produced and the section is visualized as if we were looking at a slice through the
body. This visualization enables us to analyze the soft tissue
structure and to detect abnormal or pathological conditions of the
tissue on the basis of direct information.

Physicians in clinics generally require an instructive representation and many ultrasonic imaging techniques have therefore
been developed as described by Stroke et al. (1974) and Barber
et al. (1974). Because of its accuracy, reproducibility, and
flexibililily, computer processing has been introduced in recent
years in clinical diagnosis. For example, in the field of cardiology, digital image processing provides a powerful tool for the
analysis of cardiological data. Once the ultrasonic images are
digitized, we can take full advantage of various digital image
processing techniques. Digital image processing facilitates the
reconstruction of three-dimensional structure and physical measurement of various parameters, e.g., heart volume as given by Tasto
(1974) and DeJong and Slager (1975).

The general problem of reconstructing the internal structure of a multidimensional object from a set of its two-dimensional projections has been a classic problem in physics. A number of reconstruction techniques in biology and medicine have been proposed. For example, a three-dimensional radiographic imaging system was developed by Grant (1972). However, this method is not applicable to ultrasound tomograms because the ultrasound tomogram is not a two-dimensional projection of a three-dimensional structure. Other methods of reconstruction have been used such as casts for three-dimensional computer reconstruction (Heintzen et al., 1974, 1975) and a technique using computer-reassembly and plotting of morphological serial sections in three dimensions (Willey et al., 1973). The present paper proposes a system which automatically produces a three-dimensional image of the heart of a patient by using its two-dimensional representations, i.e., its tomograms.

It is useful to have a three-dimensional display device for many applications. Plott et al. (1974, 1975) developed a three-dimensional system which produces rotated two-dimensional images and presents these images separately to the left and right eyes producing a three-dimensional display. Recently, Optical Electronics Inc. has produced analog graphic 3-D display building block modules which drive conventional xy CRT-type displays and provide three-dimensional images. Three-dimensional displays can also be produced by using a varifocal mirror (Muirhead, 1961) or a rotating mirror (Simon, 1975). Stereoscopic display and three-dimensional computer-generated movies have been produced using a rapidly vibrating varifocal mirror by both Traub (1967) and Rawson (1968).

The present authors have implemented a binocular stereoscopic display system for echocardiography on a minicomputer (Tanaka and Tamura, 1977). The present paper describes the hardware implementation of a new three-dimensional display system using an overlaid display with different gray levels corresponding to depth. The input to the system is a set of serial tomograms derived from either ultrasound or computed (x-ray) tomography (CT). Further, this system can also be used in industrial applications.

2.   SYSTEM OUTLINE

This section provides illustrations of time-synchronized echocardiograms and of the display system which we have developed.

2.1 Tomograms

Tomograms are displayed on the CRT of the ultrasonic diagnostic equipment with a few millimeters resolution using a frequency in the range of 2-3 MHz. An example of a set of tomograms obtained with this equipment is shown in Figure 1. These are longitudinal tomograms (also called B-mode or 2-D echocardiograms) which are obtained at intervals of 5mm. Figure 1(1) is taken at 30mm left from the center line of the body and Figure 1(9) is at 70mm. In order to obtain these phase-selected tomograms in a required phase of the cardiac cycle the ultrasonic apparatus operates in synchronization with the cycle.

2.2 Functions

Figures 2, 3, and 4 illustrate our 3-D display system, its block diagram, and a schematic representation, respectively. Tomograms are processed as follows.

2.2.1 The serial tomograms as shown in Figure 1 are digitized by the image input device. The digitized image is then stored on a floppy disk and can be read out at any time.

2.2.2 The 3-D display shows a 3-D image which represents its depth by its gray level, i.e., each tomogram is binarized and displayed using a gray level which corresponds to its depth. Furthermore, the 3-D image can be rotated by the image processing unit.

2.2.3 The system can synthesize a new cross-sectional image from each set of serial tomograms obtained.

2.2.4 The system can measure the volume of organs such as the heart, liver, kidney, etc.

3. SYSTEM ARCHITECTURE

Figure 5 shows the system architecture from the viewpoint of multi-microprocessor system. It is composed of a hierarchical distributed microprocessor system with loose connections via an external bus. The global processing unit (GPU) controls the local processing units (LPUs). The main features of the system are the following:

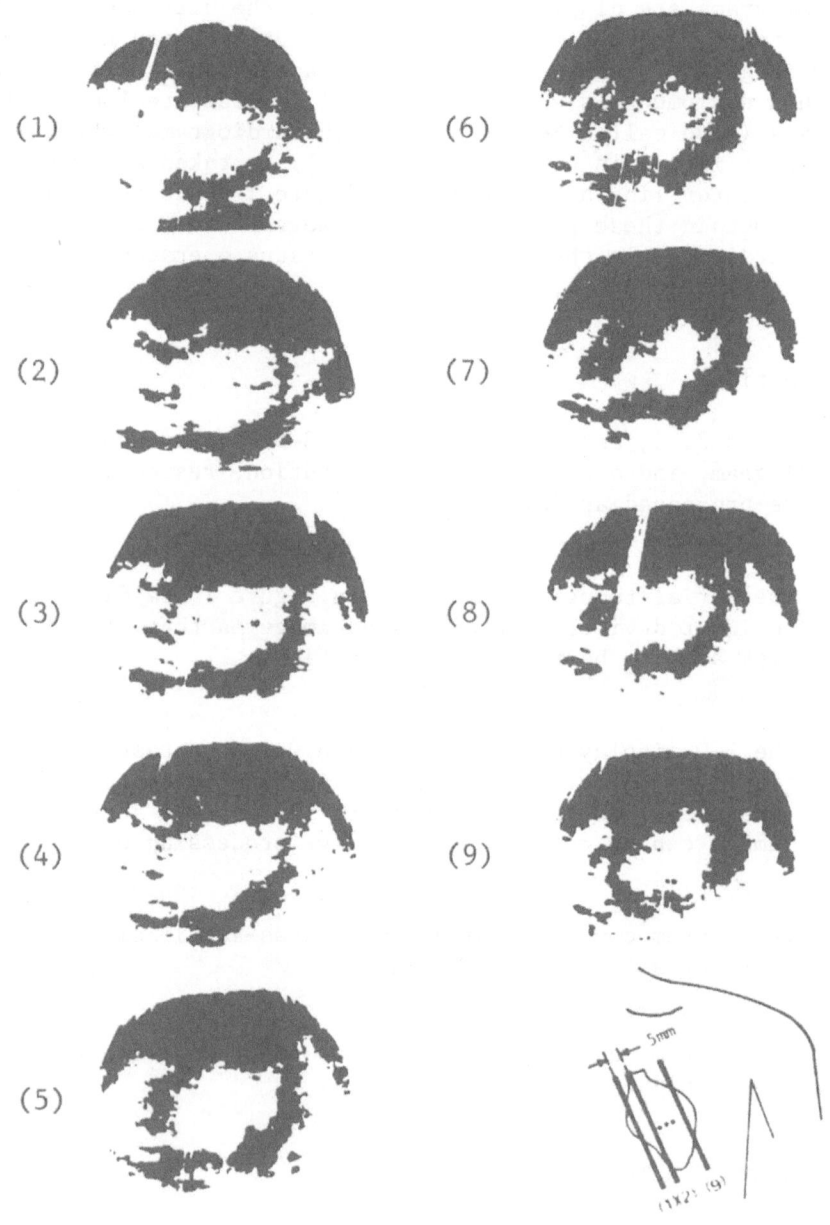

Fig. 1. Ultrasonic tomograms (healthy subject: end diastole).

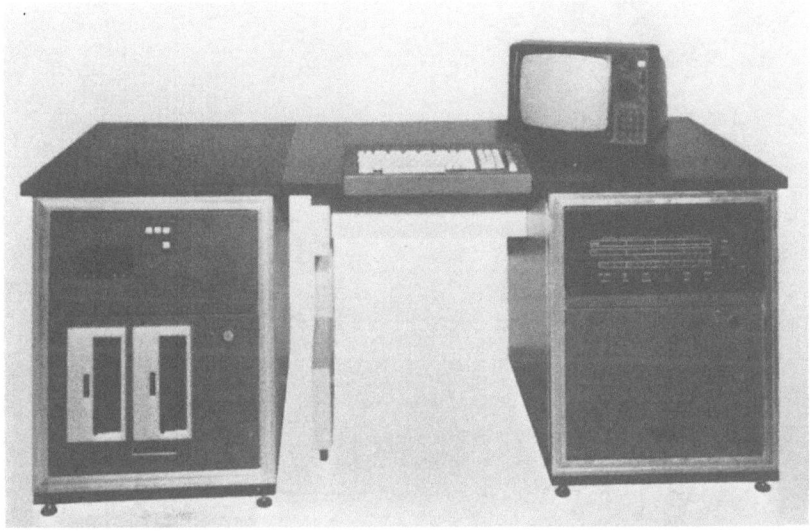

Fig. 2. Three-dimensional display system for serial tomograms.

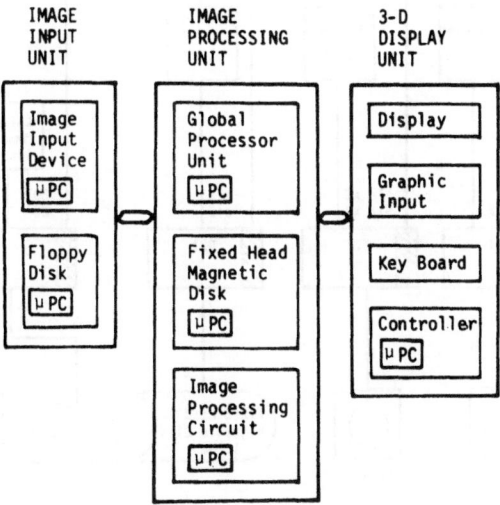

Fig. 3. System block diagram

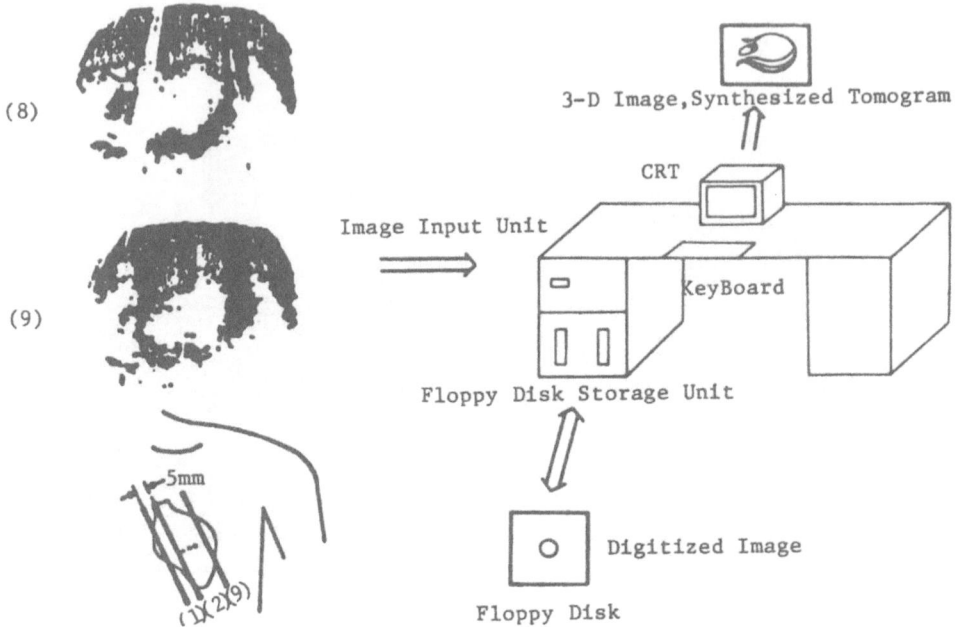

Fig. 4.   Schematic diagram of the system.

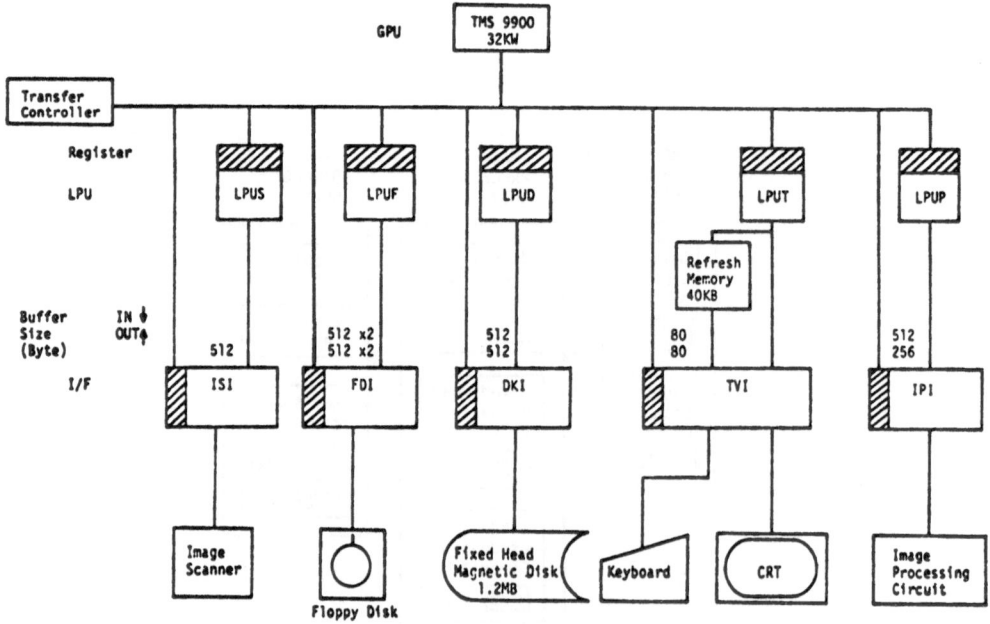

Figure 5.   System architecture.

3.1 Due to the loose connections, each device can also work as a stand-alone intelligent terminal giving the system extensibility and flexibility.

3.2 The number of data transfers between the LPUs is rather large. Consequently, we have made the hardware transfer controller capable of transferring data at high speed.

3.3 The GPU is not connected directly to the terminal devices but controls them by way of the LPUs which control each terminal device. Therefore, the GPU can control each terminal device in a common communication mode irrespective of the particular identity of each device.

3.4 By giving some LPUs their respective commands successively, each LPU works in a parallel mode. In the course of execution by some LPUs, the GPU can execute another job, for example, preparation of the next data to be processed, generation of commands to transfer the processed data, etc. Thus the load on the GPU is fairly low and, consequently, system throughput becomes fast.

The GPU uses a 16-bit microprocessor (TMS 9900) which (1) has a high performance instruction set, (2) can perform fast processing with sixteen-levels of interrupt processing, and (3) has multiplication and division functions in hardware. The LPU uses an 8-bit microprocessor (8080A) because of its tractability. The LPU performs device control, device diagnosis, and communication with the GPU by programs stored in ROM. As shown in Figure 6, the LPU has three kinds of registers, i.e., the data register, the command register, and the status register. The data register and command register may be likened to a mailbox from the GPU to the LPU. The status register is a mailbox from the LPU to the GPU. WHen data are stored in either the command register or the status register, interrupts are generated in either the LUP or the GPU, respectively. Consequently the micro-processor on the receiving side can acknowledge that the data have been stored. The block diagram of GPU-LPU communication is shown in Figure 7. Communication procedures are flow-charted in Figure 8.

4. IMAGE INPUT DEVICE

The image input device is an electro-mechanical scanner with a one-dimensional image sensor of 512 elements orthogonal to a one-dimensional mechanical drive. Therefore, it has higher positional accuracy and lower noise than a vidicon input device. An input

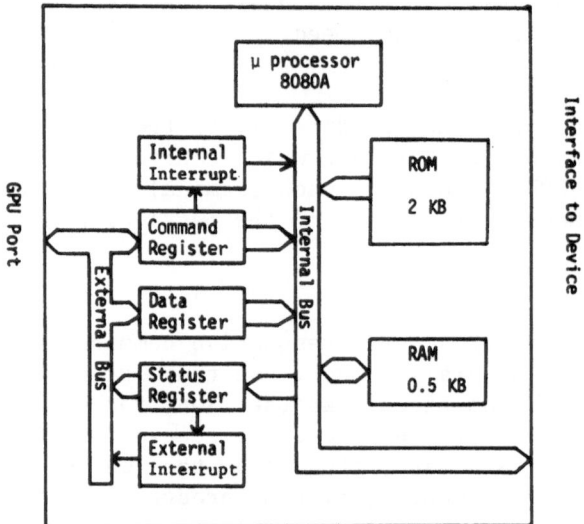

Fig. 6. Block diagram of an LPU.

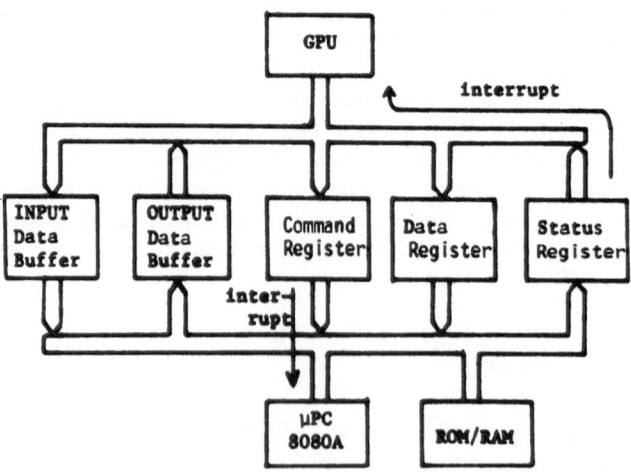

Fig. 7. CPU and LPU communication method.

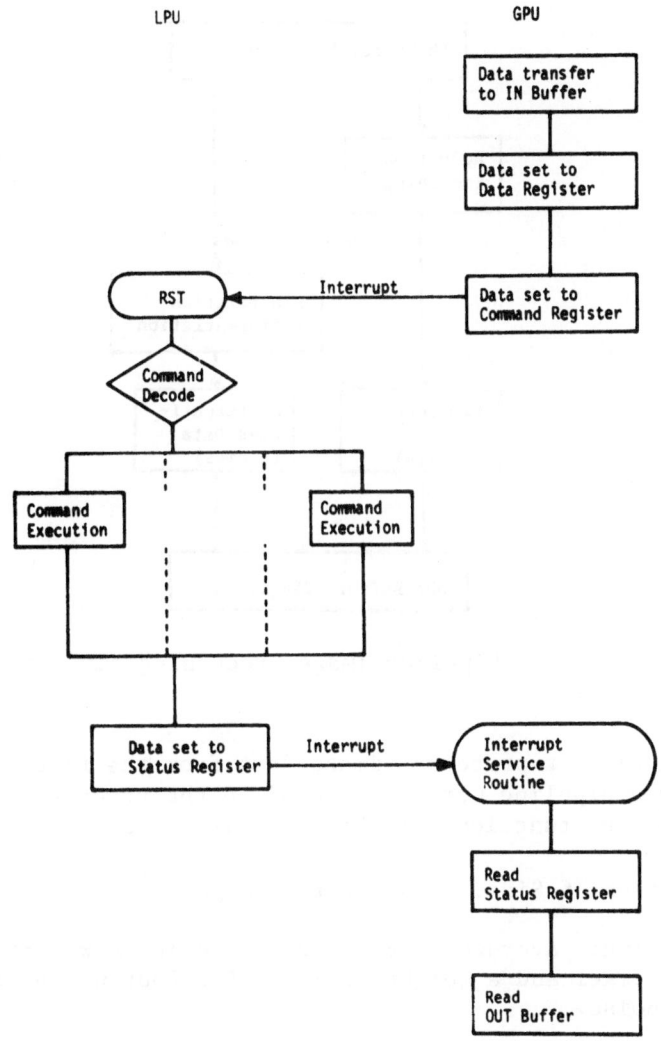

Fig. 8. Flow chart for CPU and LPU communication.

image is digitized to eight bits. The digitized data are usually stored on a floppy disk. The size of the digitized image is 512x640, which is reduced by 1/(4x4) by an image compression circuit before being stored on the floppy disk. (See Figure 9.) Since this device uses a mechanical driver in one-dimension, it can also scan long pictures, e.g., rolls of ECG paper.

5. IMAGE PROCESSING

Image data from either the input device or the floppy disk are first stored in a fixed-head magnetic disk. The image information

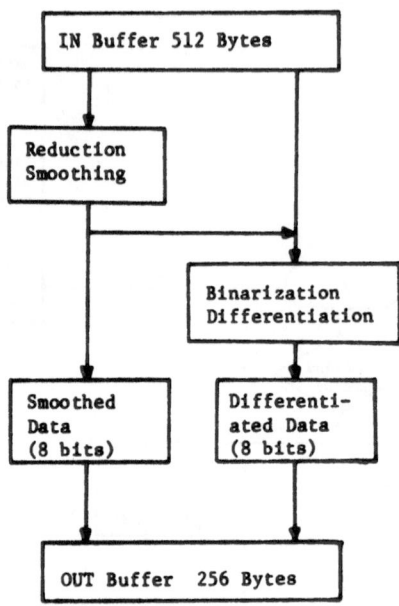

Fig. 9. Pipeline image processing circuit.

processing circuit processes image data from the fixed-head magnetic
disk using a pipeline method to speed up the processing. (See
Figure 9.) The functions of this circuit are as follows:

5.1 Reduction of Size. Reduces the image size by 1/(4x4).

5.2 Smoothing: Averages the image data using a weight of 4 for
the center pixel and a weight of 1 for the four neighbors to
eliminate noise.

5.3 Binarization: Generates binary images by the use of an appropriate threshold manually selected or automatically obtained.

5.4 Differentiation: Extracts boundary points from binary images
through a gradient operator. (See Figure 10.)

5.5 Rotation: Rotates 3-D images. The circuit for rotation is
composed of multiplication circuits and a sine-cosine function
table stored in a ROM (Figure 11). The multiplication circuit can
execute an 8-bit multiplication by (1) setting the multiplicand
and (2) the multiplier and (3) reading-out the multiplication
result. Rotation can be executed using this multiplication circuit
by three steps as follows: (1) setting of the angle of rotation,
(2) setting the x- and y-coordinate before rotation, and (3) storing
the x- and y-coordinate after rotation.

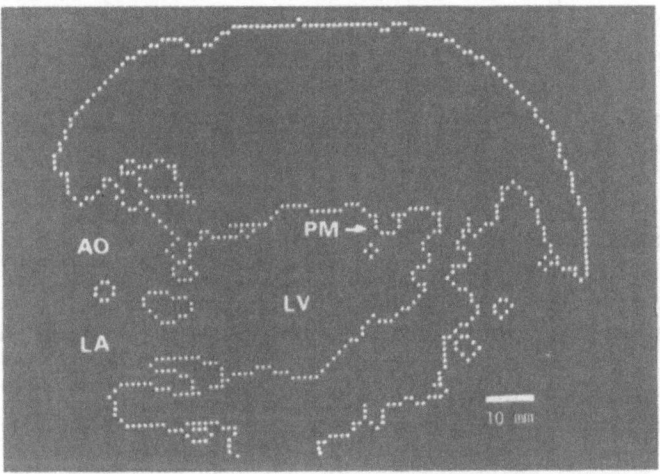

Fig. 10. Boundary points found in Figure 1(4).

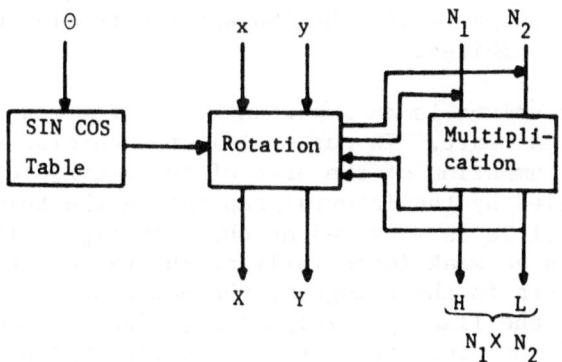

Fig. 11. Rotation and multiplication circuit.

6.   THREE-DIMENSIONAL

     The 3-D display method employed in our system is to overlay
the binary tomograms representing the depth of each tomogram by a
different gray level in accordance with its depth. Thus we can
identify the depth by the respective gray levels. This method presents a monocular 3-D image which is different from a binocular one
since it has less of a 3-D feeling。 However, this monocular method
has the advantage that it is easy to display. Also many viewers
can watch the 3-D image simultaneously.

The 3-D images are displayed on an ordinary TV screen through a VHF converter. Our device has five planes of 256x256 refresh memory and can display 32 levels of gray. Because the display device is an LPU-controlled intelligent-terminal we can edit data for write-in or read-out, we can check input data from the key board, and we can display data in the buffer.

7.   ILLUSTRATIVE EXPERIMENT

Figure 12 shows a 3-D image of a ham-phantom with a cone-shaped concavity. This image is composed of five ultrasonic tomograms. We can clearly observe that the cross section of the concavity at the center of the tomogram nearest to the observer is wider than that of the more distant tomograms.

Figure 13(a) shows a frontal view of a 3-D echocardiographic image illustrating the case of an atrial septal defect (ASD). Figure 13(b) shows the image of Figure 13(a) viewed from an angle of 20° upwards. Figure 14 shows the synthesized cross-sectional image which is perpendicular (coronal) to the sagittal pictures. Such an image cannot be generated directly and requires our synthe-sizing method. In Figure 14, the ASD and the remaining atrial septum are clearly exhibited.

The cardiac volume is an important index for measuring the pump function of a heart. In our system, the cardiac volume is measured as the summation of the area of the ventricle in each tomogram multiplied by the interval separating the tomograms. The area of the ventricle is measured as shown in Figure 15. If the echo intensity is so weak (especially at the valves) that the boundary disappears in the tomogram, the human operator adds the boundaries using the light pen (Figure 15, right) to define a closed area. If the operator specifies an internal point, then the system measures the area by propagating waves from the speci-fied internal point.

8.   CONCLUDING REMARKS

This chapter describes the hardware implementation of a 3-D imaging system using an overlaid display with different gray levels representing depth. Since we have aimed to develop a general pur-pose display, we use input off-line from the original image source. The system can deal with not only ultrasonic tomograms but also x-ray computed tomograms. We are now attempting to develop a new type of ultrasound parallel scanner and to interface it to our dis-play system to permit visualization of 3-D images on-line and in real-time. Our system will be a powerful tool not only for medical diagnosis but also for industrial inspection.

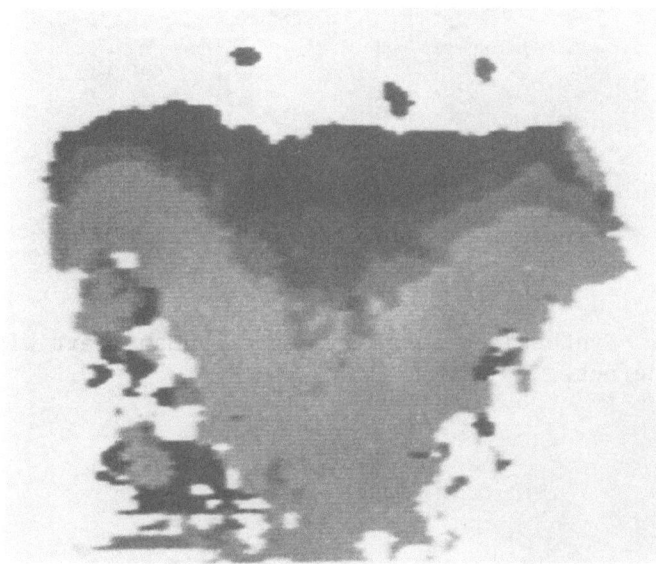

Fig. 12.   Ham phantom with cone-shaped concavity.

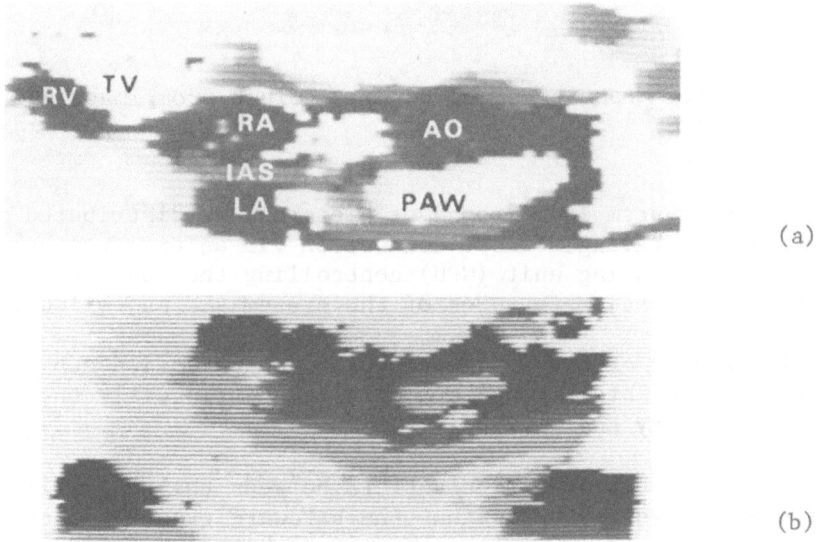

(a)

(b)

Fig. 13.   Three-dimensional images of a human heart exhibiting
atrial septal defect showing (a) frontal view and (b) view from a
20° upward angle.

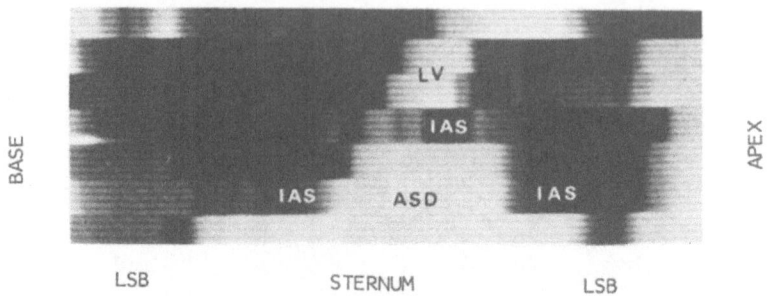

Fig. 14. Synthesized coronal view of human heart with atrial
spetum defect.

Fig. 15. Measurement of cardiac volume from image edited using
light pen.

The system is composed of hierarchical distributed microprocessors having a loose connection via an external bus with the
global processing unit (GPU) controlling the local processing units
(LPU). The major features of the system are as follows:

8.1 Because of the loose connection, each device can work as a
stand-alone intelligent terminal, and consequently the system has
extensibility and flexibility.

8.2 The number of data transfers between the LPUs will be fairly
large in this system so that the hardware transfer controller which
handles data transfer must operate at high speed.

8.3 The GPU is not connected directly with peripheral devices but
only controls them by way of the LPUs which directly control each
device. Therefore, the GPU can control each device in a common communication mode irrespective of the particular characteristics of
each device.

8.4. By giving some LPUs their respective commands successively, each LPU works in a parallel mode. In the course of execution of some LPUs, the GPU can execute another job, for example, preparation of data to be processed, generation of commands to transfer the processed data, etc. Thus the load of the GPU is comparatively low and consequently system throughput is fast.

8.5 The image data from either the input device or the floppy disk are first stored on a fixed-head magnetic disk. Next the image processing circuit processes the image data on the fixed-end head magnetic disk by pipeline methods to achieve high processing speed.

9. ACKNOWLEDGEMENTS

This work was supported by a grant from the Foundation for Research on Medical and Biological Engineering. The display system was developed through the cooperative work of the members of the research committee of the Foundation for Research on Medical and Biological Engineering, Prof. H. Abe, Lecturers M. Inoue and H. Matsuo, Dr. M. Matsumoto of the Osaka University Hospital, Dr. T. Nakahara and Dr. K. Yoshida of Sumitomo Electric Industries, Ltd. Further, the authors would like to thank J. Hiramoto and Mr. K. Hirano of Sumitomo Electric Industries, Ltd. for their help in producing this system.

10. REFERENCES

Barber, F. E., Baker, D. W., Nation, A. W. C. et al., "Ultrasonic Duplex Echo-Doppler Scanner," IEEE Trans. Biomed. Eng. BME-21: 109-113 (1974).

de Jong, L. P. and Slager, C. J., "Automatic Detection of the Left Ventricular outline in Angiographs Using Television Signal Processing Techniques," IEEE Trans. Biomed. Eng. BME-22:230-237 (1975).

Grant, D. G., "Tomosynthesis--A Three-Dimensional Radiographis Imaging Technique," IEEE Trans. Biomed. Eng. BME-19:20-28 (1972).

Heintzen, P. H., Brennecke, P., Bürsch, J. H., and Lange, P., "Automated Video-Angiocardiographic Image Analysis,"

Heintzen, P. H., Moldenhauer, K., and Lange, P. E., "Three-Dimensional Computerized Construction Pattern Analysis," Europ. J. Cardiol 1:229-239 (1974).

Muirhead, J. C., "Variable Focal Length Mirrors," Rev. Sci.
    Instr. 32:210-211 (1961).

Plott, H. H., Irwin, J. D., and Pinson, L. J. "A Real-Time Stereo-
    scopic Small-Computer Graphics Display System," IEEE Trans.
    Syst. Man. Cybern. SMC-5:527-533 (1975).

Plott, H. H., Irwin, J. D., and Pinson, L. J., "A Three-Dimensional
    Display System," IEEE Trans. Educ. E-17:152-157 (1974).

Rawson, E. G., "Three-Dimensional Computer-Generated Movies Using
    A Varifocal Mirro," Appl. Opt. 7(8):1505-1511 (1968).

Simon, W., "A Three Dimensional Computer Display," Compt. Graphics
    Image Process. 4:396-402 (1975).

Stroke, G. W., Kock, W. E., Kikuchi, Y., and Tsujiuchi, J. (eds.),
    Ultrasonic Imaging and Holography-Medical, Sonar and Optical
    Applications, New York, Plenum Press (1974).

Tanaka, K. and Tamura, S., "Binocular Stereoscopic Display System
    for Echocardiography and Computer Synthesis of Tomograms,"
    Proc. 2nd World Conf. Med. Inform. MEDINFO77, Toronto (1977),
    pp. 999-1004.

Tasto, M., "Motion Extraction for Left-Ventricular Volume Measure-
    ment," IEEE Trans. Biomed. Eng. BME-21:207-213 (1974).

Traub, A. C., "Stereoscopic Display Using Rapid Verifocal Mirror
    Oscillations," Appl. Opt. 6(6):1085-1087 (1967).

Willey, T. J., Schultz, R. L., and Gott, A. H., "Computer Graphics
    in Three Dimensions for Perspective Reconstruction of Brain
    Ultrastructure," IEEE Trans. Biomed. Eng. BME-20:288-291
    (1973).

DYNAMIC COMPUTED TOMOGRAPHY FOR THE HEART

Y. Tateno, Y. Umegaki, G. Uchiyama and Y. Masuda

National Institute of Radiological Sciences and
Chiba University Hospital
Chiba, JAPAN

1.   INTRODUCTION

Real-time computed tomography (CT) of the heart is a subject of great potential importance (Iinuma, 1977). Real-time images of the cardiac ventricles could be used to evaluate stroke volume (Miller, 1977) and ejection fraction. Real-time images of the heart muscle could be used to detect damaged tissues and, presumably, show edematous and necrotic regions (Adams, 1976). Moreover, sufficiently fast scanners in combination with contrast media injections could enable one to get time-concentration curves in the myocardium which provide a measure of regional perfusion.

There are two possible ways to realize these potentialities: (1) an ultrafast CT scanner and (2) electrocardiogram (ECG) synchronized CT. This chapter treats several kinds of real-time cardiac CT using a computer controlled flying spot x-ray microbeam generator.

2.   DYNAMIC SCANNER

A CT scanner which we developed at Chiba University Hospital (Tateno, 1976) can be gated in order to synthesize static images of the systolic and diastolic phases from a number of successive heart beats. The machine is called the "Dynamic Scanner." The most remarkable feature of the Dynamic Scanner is the x-ray generator. The generator's appearance greatly differs from that of a conventional x-ray tube. The electron gun is composed of a point-source cathode and a Wehnelt cylinder. The electron beam emitted from the source is accelerated by voltages of up to 140KV, focused

by the electromagnetic objective lens, drawn in the axial direction, brought into the optical axis, and finally projected on to a
small spot on the target. Before hitting the target the electron
beam is rapidly deflected with high accuracy by a deflection coil
in accordance with positional instructions from a scan generator.
An intense microfocused x-ray beam is emitted from the target.
After passing through a pinhole lens, the x-ray beam becomes a
microbeam or a fan beam which can be used to irradiate any desired
portion of the body because the x-ray beam can be directed under
control of the electron beam.

The Dynamic Scanner provides many kinds of images: (1) Scanning x-ray images at ultra-low dosage, (2) x-ray contour tracing
under on-line computer control, (3) x-ray tracing of a rapidly
moving object by on-line computer, and (4) CT images.

2.1   CT Images Using the Dynamic Scanner

The principle of making CT images by this instrument is as
follows. The electron beam of the x-ray generator is scanned over
the target so that the x-ray microbeam will traverse a cross section of the patient's body. The x-ray beam is collimated by a
pin-hole lens so that it becomes a microbeam.

A long sodium iodide detector detects the transmitted beam
intensity when the x-ray beam hits the detector. It does not need
to detect the position information since position is already determined by the scan generator under computer control. The whole
system including the x-ray generator and the long detector rotates
around the patient. The microbeam traverses the body section over
a certain interval of the rotational angle. All of these operations are performed under computer control.

2.2   ECG Gated Image Using the Dynamic Scanner

The Dynamic Scanner can quite easily be synchronized with the
ECG. The x-ray microbeam can be generated for 50ms triggered by
signals derived from the R waves of the ECG. Thus one can obtain
a cross-section CT image at a certain phase of the cardiac cycle.
Data are acquired at each degree of rotation in step-by-step mode
with a total rotation angle of 230 degrees. The patient is told
to hold his breath at maximum inspiration while the machine is
working. The gantry rotation and x-ray radiation stop when the
patient needs to breathe. The overall time necessary for obtaining a gated image of one phase is 6 to 10 minutes.

In one study, two tomographic section images of the heart in
the end diastolic and end systolic phases were obtained by setting
the gates at the peak of the R wave and at the middle of the down
slope of the T wave, respectively. At end systole, contracted
heart and enlarged pulmonary vessels were seen. In images of
cardiac infarction, changes in the size and shape of the heart are
less clear than in normal cases. More than 50 cases covering
several heart diseases have so far been examined. In order to
extract information on cardiac wall movement, subtraction techniques have been applied.

2.3 ECG Phase Differentiation Method

Since the time to generate the ECG gated image is fairly long,
for diseased people we tried to minimize the data acquisition time
and developed another method of ECG synchronization. This method
is called the ECG data-sorting method or phase differentiation
method. After data for all 230 views per slice were acquired in
11 seconds along with the ECG signals, data sorting was made
according to the segments of ECG cycle. Four to 8 scans of the
same body section are required to obtain the inputs that cover all
segments of a cardiac cycle. By this method, any segment of a
cardiac cycle can be reconstruction at any time after the data
acquisition is completed.

Figure 1 shows 23 cardiac CT images corresponding to 23 segments of an ECG cycle, taken by this method. From these images,
the changes of cardiac area can be measured. Clinical data have
been calculated and are shown in Figures 2 and 3. These data may
be used to evaluate stroke volume and ejection fraction.

2.4 Comparison of Methods

The two different ECG synchronization methods which we use
presently are summarized as follows. The ECG gated method uses a
fan beam single scan geometry and provides a single phase from
each single scan. The phase differentiation method uses a fan
beam multiscan and gives multiphases from each single scan.

Making a comparison between these two methods, one could point
out that the former gives perfectly synchronized images, but requires a larger number of heart beats and viewing angles; the latter
is able to produce almost synchronized images within a shorter time.

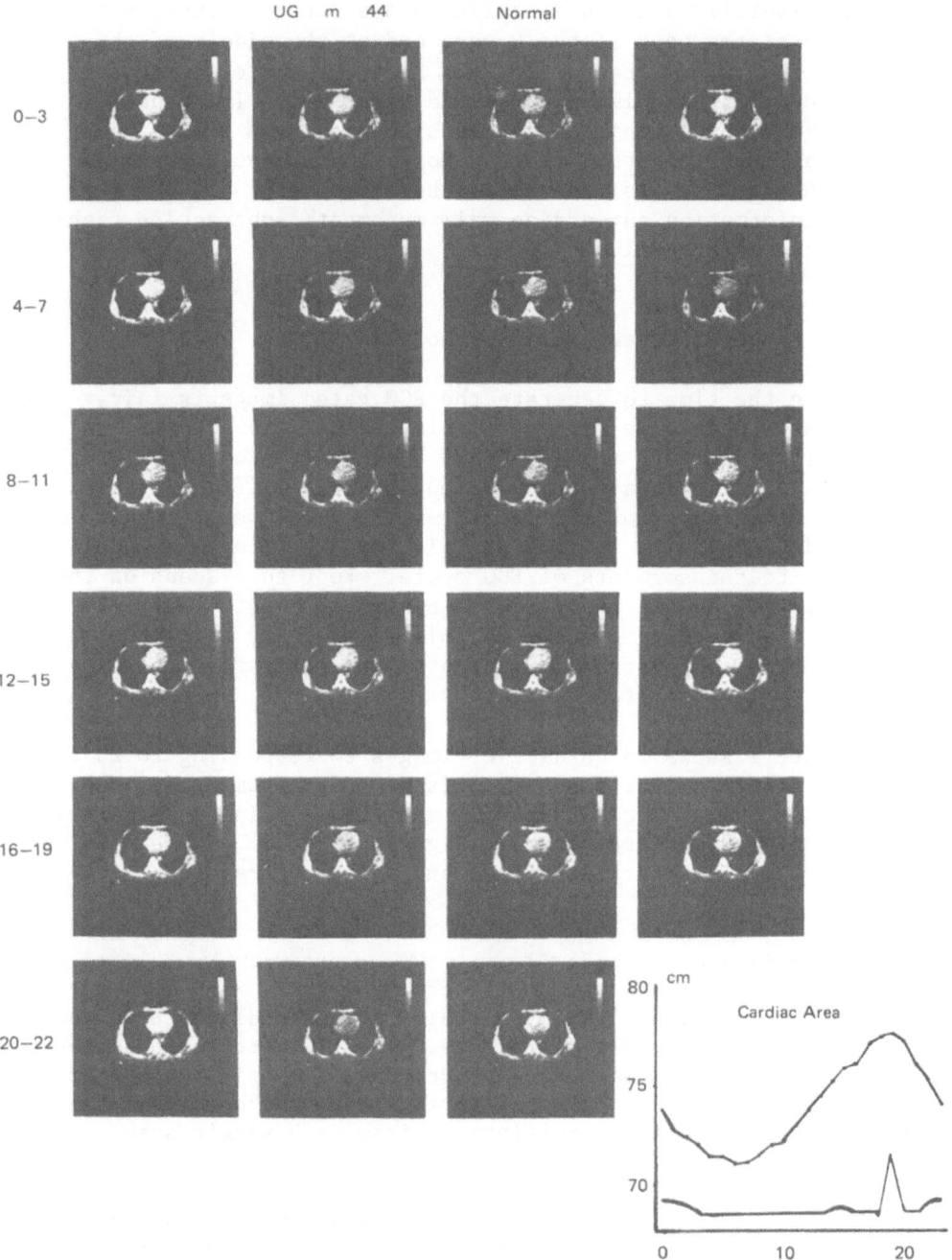

Fig. 1.   Cardiac CT images corresponding to 23 segments of an ECG cycle.

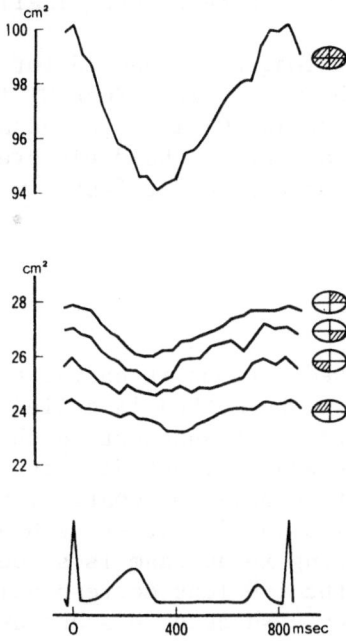

Fig. 2.   Changes of cardiac area of a normal heart.

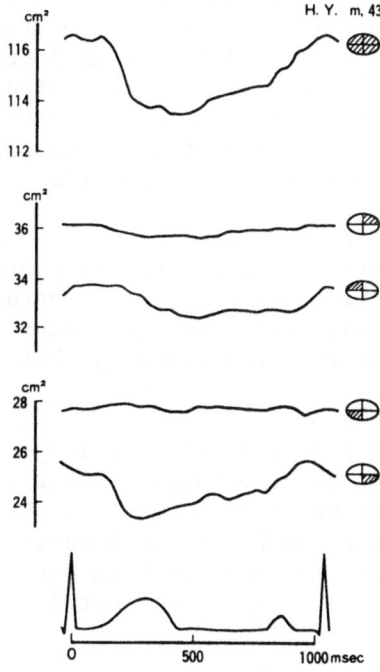

Fig. 3.   Changes of cardiac area for a case of extensive anterior infarction.

3.   PROPOSED METHOD FOR FASTER SYNCHRONIZATION

Higher quality CT images are needed for the detection of
damaged areas in the heart muscle. Such CT images require at
least 120 views. In cardiac CT it is necessary to obtain them
within a time that a patient can hold his breath. For the purpose
of overcoming these problems, other faster methods for ECG synchronization are proposed.

3.1 Multilens Method

One of the proposed methods is characterized by lens-switching
of the scanning x-ray. The method is called the multilens method
(Figure 4). The generator arrangement in this method is similar
to that of the Dynamic Scanner, but differs from it as to the wider
range of x-ray generating position control, possession of a multiple
lens and a rotating high speed shutter mechanism which cuts the
x-ray beam. The scanning x-ray beam is successively emitted from
the 1st lens through the ith lens corresponding to the ECG synchronized gate signal. These successive scans are made withon 0.2
second.

Each lens is placed at an interval that corresponds to the
desired angle. Thus data can be obtained in a series of regular
intervals. The scanner is continuously rotated at high speed and
in such a manner that the $(i+1)^{th}$ view projection is made by the
1st lens after the ith x-ray scan is completed by the ith lens,
thus covering the width of the patient at a sufficient fan angle.
The next block scan is executed at the time when the next ECG-
gated signal of the same phase is produced.

In this manner, the lens shutter must be open at the desired
time and desired position so that the x-ray projection is achieved
at the corresponding time and position. Therefore, the rotation
speed must be finely adjusted. The generator is rotated at a
speed roughly equal to the mean velocity that relates to the
patient's heartbeats.

As was explained above, cardiac CT images of specific phases
of the heart's motion can be obtained when the patient holds his
breath while his heart beats 20 times. Furthermore, different
phase cardiac CT images can be obtained during one rotation period,
provided the instrument is equipped with data-storing buffers,
since different phase data can be collected at each spacing interval.

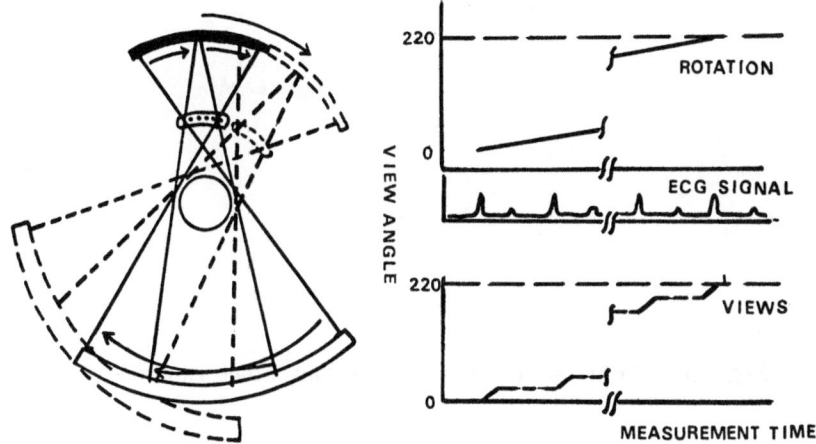

Fig. 4.   Schematic of multilens method.

3.2   Rocking Fan Beam Method

The other proposed method is somewhat similar to the first one as to the high speed x-ray generation control by means of electronic control of the electron beam. However, it is not a true x-ray microbeam scan, but rather a multi-view projection of fan-shaped x-rays. Its x-ray generating point is somewhat larger than that of the first method. This method is called a rocking fan beam. With this method, the scanner is equipped with multiple detectors and performs multiple view projections at high speed (Figure 5).

Using this method, one x-ray absorption datum is gathered within a few milliseconds, using either scintillation crystals or semiconductor multidetectors. The measurement time in this method is only restricted by the cycle time, which consists of the time for a measurement of the required data, the time for storing them in the buffer storage, and the time for returning to stand-by for the next measurement.

In the interval between measurements, x-rays are cut by electronic shuttering. Thus, successive views are gathered in the same manner as the first method, requiring only a few milliseconds for each. The x-ray generator continues to rotate during this time interval in order to prepare the next block of x-ray projections. For example the $(i+1)^{th}$ projection is made from the 1st x-ray generating point on the generator target in the previous 1st block projection.

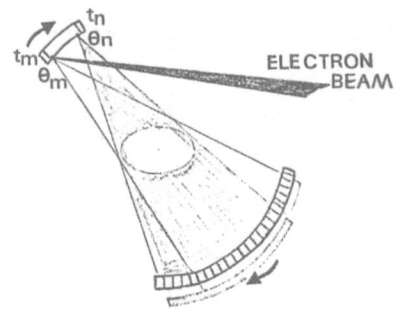

Fig. 5.   Schematic of rocking fan beam method.

4.   CONCLUSION

The ideal cardiac CT is, of course, real time CT. We proposed and developed previously a system of ultrafast realtime CT
scanning which uses a flying-spot x-ray microbeam generator. However, considering the present state of real-time CT techniques,
the methods introduced here can be used in practice for the diagnosis of cardiac diseases until the time when true real-time CT
will become available for common use.

5.   REFERENCES

Adams, D. F., "Computed Tomography of the Normal and Infarcted
    Myocardium" Am. J. Roentgenol. 126:786 (1976).

Iinuma, T. A., "Proposed Systems for Ultrafast Computed Tomography,"
    J. Computer-Assisted Tomography 1:494 (1977).

Miller, S. W., "Right and Left Ventricular Volumes and Wall
    Measurements: Determination by Computed Tomography in
    Arrested Canine Hearts," Am. J. Roentgenol. 129:257 (1977).

Tateno, Y., "Low-Dosage X-Ray Imaging System Employing Flying Spot
    X-Ray Microbeam," Radiol. 121:189 (1976).

EFFICIENT ANALYSIS OF DYNAMIC IMAGES USING PLANS

S. Tsuji and M. Yachida

Department of Control Engineering, Osaka University

Toyonaka, Osaka 560, JAPAN

1.   INTRODUCTION

A wide variety of processes which are observed in experiments in medicine and biology are recorded on video tape or cine film as dynamic (moving) images. Analysis of these images, however, is a tedious task for human beings who must screen a large number of frames to discover the patterns of pertinent features and measure their parameters. In recent years, computer analysis of moving images, such as motion analysis of the heart as reported by Kaneko and Mancini (1973) and by Jong and Slager (1974) or of micro organisms (Greeves, 1975, and Ariki et al., 1977) has been investigated for the purpose of reducing human labor and extracting more reliable information.

Because of the large amount of information contained in a moving image, too much computing time is spent if large areas of consecutive frames are examined in detail. Although a special parallel-processor system could significantly reduce the computing time for preprocessing, feature extraction, or change detection, this might waste computing power in unnecessary processing. The analysis of moving images does not require that all information in all the records be processed. Only a small fraction need be utilized. For example, only small subareas covering moving objects are needed for tracking when motions are small. Abrupt changes from frame to frame require more carefully examination. Most motion analyzers have been designed to save computation time by utilizing models of the objects in already analyzed frames to predict their locations or shapes in the current frame (Chow and Aggrawal, 1977).

We have extended these ideas to organize a new system, called
a "plan-guided" analyzer for moving images. The novel feature of
our system is to fully utilize both *a priori* and acquired knowledge
about the dynamic image so as to extract significant and useful
information from each record at an extremely low computation cost.
A priori knowledge contains general properties such as, "The
moving objects are dark and blob-like," which is useful to design
efficient procedures. However, *a priori* knowledge lacks quantitative information such as object location, velocity, brightness
level, and shape.

The plan-guided analyzer examines a small number of low-
resolution segments of each record to make a coarse model or
"sketch" of the process from which it estimates which frames and
what regions in them are worth examining in detail, or what type
of feature extractors are likely to be most effective at each
point in each of these regions. Thus the system makes a plan,
specifying (1) frames to be sampled, (2) regions to be examined,
and (3) effective feature extractors to be applied. Since more
accurate knowledge is acquired as a result of this plan-guided
analysis, the coarse model is replaced with a fine and more reliable one, describing exact properties of the objects in the
current frame, which is then used to make a more effective plan
of sampling and analyzing the next frames. By iterating this
process, planning, analyzing, and updating, the system constructs
a final model of the process, containing all important information
required.

The purpose of utilizing the plan is not only to minimize the
computation cost but also to increase the reliability of the
extracted information. Thus, the plan selects feature extractors
and their parameters which are considered to be the most reliable
at each point. However, it is not easy to find the correct
feature at every point in a noisy dynamic image. Montanari (1971)
and Martelli (1976) developed optimization methods known to be
useful in searching noisy pictures for contours or regions having
specified properties. These methods can be applied to dynamic
images by embedding the properties of each object located in a
figure of merit. Sometimes this optimization method requires too
much memory space and computing time, especially when a global
shape property such as smoothness over a wide range is specified,
since we are forced to solve a high-dimensional, non-serial optimization problem. We, therefore, propose an improved optimization
method which utilizes an efficient search for the optimal sequence
of linear segments instead of the already known inefficient search
for the optimal series of pixels.

2.   HARDWARE SYSTEM

Our system (Figure 1) is designed to accept two types of recording media, video tape and cine film. Both have merits and demerits. We have developed two types of plan-guided analyzers, which adapt to the properties of moving objects and of the recording medium used. The application discussed in this chapter is the analysis of the spatial and temporal patterns of the external and internal surfaces of the heart as displayed in cine-angiograms. Other applications have been described by Yachida (1978) and Yachida et al. (1978).

The computer used is a HP-2108A, which is a 16 bit machine with 32K words of memory, a disc (15M bytes), a printer/plotter, a storage tube display, and three terminals. A multi-user operating system is employed. A 256K byte picture memory with an average access time of 400ns per byte is connected to the computer. We utilize this memory as a flexible buffer for video input and output in conjunction with a VTR (video tape recorder), a TV camera, and a color display. The memory is also used as a high-speed, random-access file for dynamic images using a software subsystem which swaps picture data in a segment of the memory for data on the disc on a first-in-first-out basis. Our system is also provided with a computer-controlled film transport which selects a specified frame by moving 35mm film at a rate of 15 frames/sec. At present, the TV camera is used as the input transducer for film images which are transformed into a 256x256 6 bit digital picture.

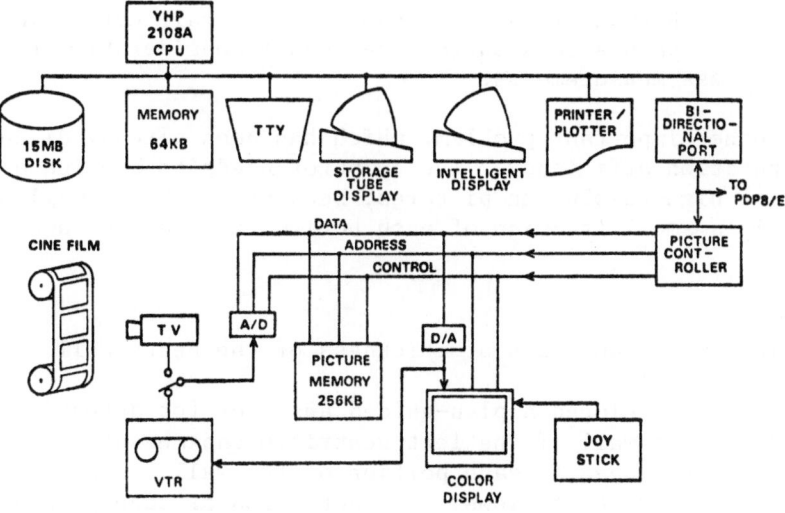

Fig. 1. Block diagram of the hardware system.

3.   ANALYSIS OF HEART WALL MOTION IN CINE-ANGIOGRAMS

In recent years, people have studied computer analysis of the
cine-angiogram which is a cine film record of the radiographic
image of a beating heart into which x-ray opaque dye is inter-
mittently injected through a catheter. The primary goal is to
measure the temporal change in the volume of the heart. The computer is used to detect the internal surfaces of the heart in each
frame and then calculates heart volume by utilizing a three dimensional model of the heart. In this chapter, a more advanced
analysis is described, which detects both internal and external
surfaces of the heart, and measures the spatial and temporal change
in heart wall thickness. This gives us important new measures of
heart disease.

The usual method of detecting the boundary of the left
ventricle uses a rather simple procedure; for example, an adaptive
threshold is employed to detect candidate points for the boundary
and then a simple statistical method is used to examine their
neighbors and fit line segments to the boundary. This is the
approach taken by Kaneko and Mancini (1973). One reason why this
detection method works is that the cine film is taken and developed
so as to show a rather sharp contrast in gray values between the
light ventricle and the dark background. Such a photographic
technique, however, is not applicable to cine-angiograms taken for
use in analyzing heart wall thickness since intermediate gray
values are required in such radiographic images. As a result,
these images contain much noise, masking the small gray value
changes on or near the heart wall surfaces. Yet an accurate
detection of both surfaces is needed, since small errors in the
locations of points on a surface can cause considerable errors in
the thickness measurements.

Another important problem, which has been already discussed,
is computation efficiency. The computer needs to examine a large
number of high-resolution pictures, because the heart wall occu-
pies only a small fraction of each horizontal scan in each input
picture.

3.1 Plan-Guided Analysis of Thickness of the Heart Wall

We have developed a plan-guided analyzer for detecting both
surfaces of the wall of the left ventricle and measuring temporal
changes in thickness at each portion of the wall. The features
of the system are as follows: (1) detection of edge points in
the first frame is guided by a plan which specifies regions to be
examined and promising feature extractors to be used, (2) smooth
boundaries of the wall are obtained by an efficient search method,

and (3) another plan is utilized to analyze consecutive frames; it selects the next frame to be sampled and guides the analyzer so as to examine this frame in a much more efficient manner than the first frame.

The analysis is divided into the following five phases:

(1) Planning of feature extraction for the first frame.
(2) Search for smooth boundaries.
(3) Selection of the next frame.
(4) Analysis of consecutive frames.
(5) Measurement of thickness.

3.2 Input Pictures

Input to the system is a sequence of x-ray images of the left ventricular chamber recorded on 35mm cine film at a rate of 60 frames/sec; thus, a record of one cardial cycle contains about 50 frames. A frame is converted into a 256x256 6 bit digital picture. Figure 2 shows a typical frame and an example of the result of applying a simple 3x3 gradient operator to one frame. The result contains much noise, especially significant is the non-uniformity of gray values in the ventricular chamber near the internal surface, where images of the muscles and tissues attached to the surface generate noise. The contrast of gray values from one side of the heart wall to the other is very low, with gray values in the wall gradually increasing, as the mean, with distance from the external surface.

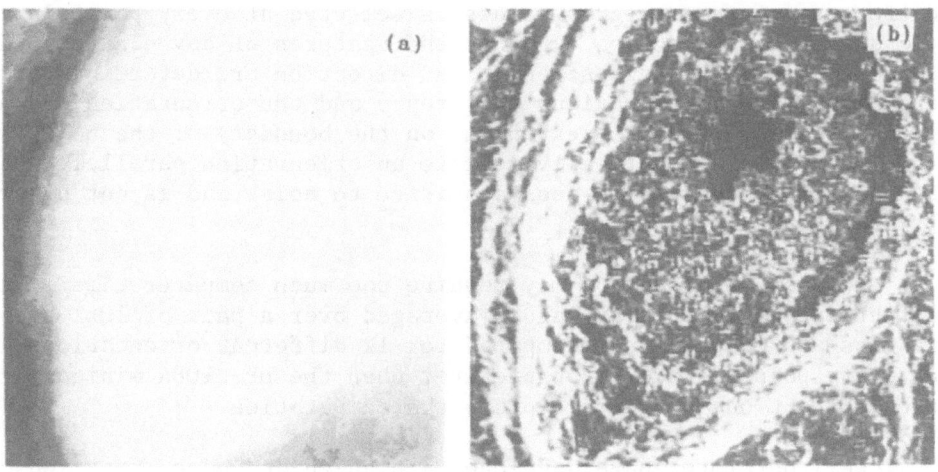

Fig. 2. A. Typical input picture from a cine-angiographic recording. B. The result of applying a 3x3 gradient operator.

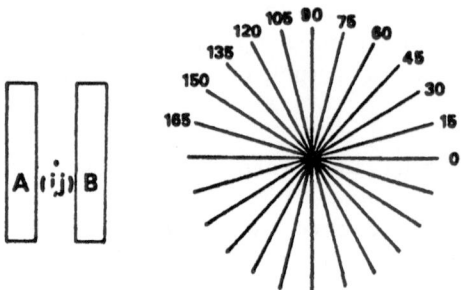

Fig. 3. Paired rectangles used in the detection of edge strength
and orientation.

3.3 Planning for Feature Extraction

Because of the low picture quality of a cine-angiogram, such
as low contrast and much noise, a level detecting method cannot
reliably find a heart wall. As shown in Section 3.2, if we apply
a gradient operator to a small area around every point in the picture, then the noise generated by the operator masks the boundaries
of the wall. In order to reliably detect edge points in a noisy
picture, edge features should be extracted from gray values averaged
over some larger window. A window consisting of paired rectangles
of variable orientation (Figure 3) is useful as the area over which
to average gray values. The orientation which provides the largest
difference of average gray values is selected at every point in the
picture. Each selection extracts the features of any edge existing
at that point; the edge strength and direction are determined from
the already computed maximum difference and the orientation of the
paired rectangles. At every point on the boundary of the heart wall,
this procedure automatically selects an orientation parallel to the
boundary. Its output is less sensitive to noise and is not blurred
by averaging.

However, this method may require too much computer time. If
differences between gray values averaged over a pair of 10x1 rectangles are computed and compared for 12 different orientations,
at every point in a 256x256 picture, then the HP-2108A minicomputer
spends about one hour to complete the computation.

We, therefore, have developed a plan-guided edge detection
method to shorten the computation time. The system first com-
presses the 256x256 original picture into a 64x64 picture, of which
each element has a gray value equal to the sum of those in each

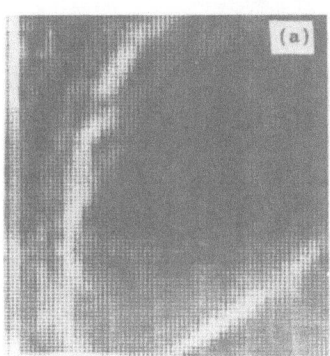

Fig. 4. a. The result of applying a 3x3 gradient operator to a compressed version of Fig. 2A. b. Analysis regions and edge orientation vectors.

4x4 window in the original picture (Figure 4a). Applying a simple 3x3 gradient operator to every point in the compressed picture and thresholding it, the system decides upon regions to be examined further and generates an approximate edge orientation vector at each point in them (Figure 4b). As can be seen, Figure 4a is less noisy than Figure 2B. Figure 4b is described as a map containing the analysis region and the directions of edges. The analyzer utilizes this map to efficiently find features; an edge feature at each point in the analysis regions is detected by a pair of rectangular areas perpendicular to the direction. Since the analysis regions occupy 1/3 of the input picture (we could considerably reduce them by using a higher threshold value or utilizing more heuristics) and since only one selected edge direction out of 12 possibilities is examined at each point, the computation time for edge finding by this method is $(1/3) \times (1/12) = 1/32$ of that without the plan. The gradient operation at each point in the compressed image is iterated about 4000 times which is 1/16 of that required for the original picture. The operation is much simpler, therefore its computation cost is less expensive. The result of applying the plan-guided edge detecting method is shown in Figure 5.

3.4 Efficient Heuristic Search for Smooth Boundaries

The next phase after edge detection is to recover the outer and inner boundaries of the heart wall in the analysis regions. Since the edge picture still contains considerable noise, and since precise detection of the boundaries is needed, classical curve followers are not applicable. Optimization methods for finding a curve of specified properties in a noisy picture have been developed which embed the properties of the edge into a

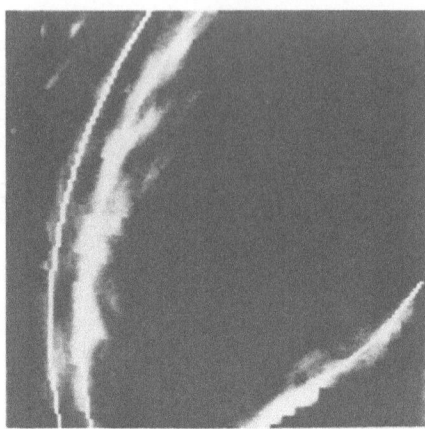

Fig. 5. Result of the plan-guided system for boundary detection.

figure of merit and search for the curve as that sequence of consecutive points which minimize the figure of merit using either
the dynamic programming technique of Montanari (1971) or the
heuristic search method of Martelli (1976)

In the analysis of the heart wall, we need to find smooth
curves over a wide distance, since local or global unevenness of
the detected curve is considered as statistically meaningless. In
order to find a smooth curve hidden in noise, the methods mentioned
above embed the property of smoothness into part of the figure of
merit as the sum of curvatures at each point. If the smoothness
over a wide range is embedded, then we are forced to solve a high-
dimensional, non-serial optimization problem, which requires too
much computing time and memory space.

A different approach has been studied to avoid this difficulty.
Since the curve to be detected is smooth, we can approximate it
with a sequence of linear segments. We embed the smoothness over
a wide range of a sequence of linear segments connecting points
$x_1, x_2, \ldots, x_n$ into a figure of merit $C(x_1, x_2 \ldots, x_n)$ given by

$$C(x_1, x_2, \ldots, x_n) = \sum_{i=1}^{n} e(x_i) + k_1 \sum_{i=1}^{n-1} |\theta(x_{i-1}, x_i) - \theta(x_i, x_{i+1})| \quad (1)$$

where $e(x_m, x_n)$ is the mean of edge strength along the segment connecting x_m and x_n, $\theta(x_m, x_n)$ is the inclination angle of the segment,
and k_1 is a constant. We apply this method to the detection of the
wall boundaries in the edge picture. The starting points of the

search are decided from *a priori* knowledge about the boundaries,
e.g., both boundaries have directions between 0° and 90° at the
upper part of the ventricle. A sequence of 10 successive edge
points on a ridge within this angular range is selected as a
starting point.

We have applied a heuristic search to find a solution to this
optimization problem. Results were successful in the case of many
input pictures. The search, however, sometimes fails, when the
outer and inner boundaries are close together at the end of the
expansion period of the ventricle. Since the edge strength on the
inner boundary is much larger than that of the outer one, a search
for the external surface sometimes crosses the wall and follows the
inner boundary in error.

In order to avoid the above difficulty, we utilize knowledge
other than the fact that gray values on the outer boundary are
darker than those on the inner one. We use a figure of merit
evaluating edge strength, smoothness, and gray values, as follows.

$$C(x_1,\ldots,x_n) = \sum_{i=1}^{n} e(x_i) + k_1 \sum_{i=1}^{n-1} |\theta(x_{i-1},x_i)$$

$$- \theta(x_i,x_i-1)| + k_2 \sum_{i=1}^{n} |g(x_i) - G| \qquad (2)$$

where $g(x_m,x_n)$ is the mean gray value of points on the line segment
connecting x_m and x_n, G is a mean gray value of a group of candidates for starting points, and k_1,k_2 are constants. Figure 5 is
an example of a result obtained by this method.

Next the detected boundaries are arranged into a model which
guides the selection and analysis of consecutive frames as
described in the next section.

3.5 Selection of Frames for Analysis

In order to reduce computation time for subsequent frames, we
utilize the model obtained from the result of the previous analysis.
First of all, narrow analysis regions are selected on both sides of
the detected boundaries (Figure 6). At present, the width of the
regions is set at a length of 7 pixels. We can save computation
time by examining picture data in these regions only.

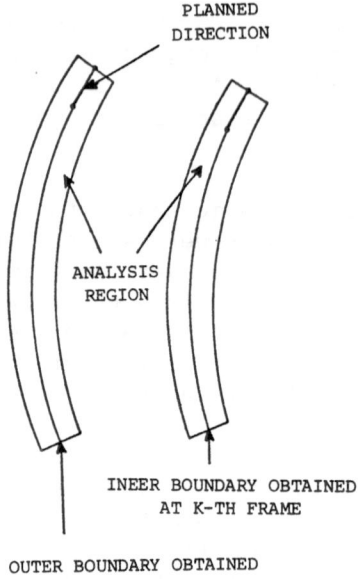

Fig. 6. Analysis regions for consecutive frames

As mentioned before, the system should select frames worth
examining from a great number of frames in the cine-angiographic
recording. Because movement of the heart is intermittent, it is
reasonable to assume that the analyzer should skip an examination
of consecutive frames in which there is little change from the
previously analyzed frame. We use a simple temporal difference
method to measure the degrees of change. If the wall boundaries
move to some extent, then there exist significant changes in the
analysis regions. For each subsequent frame we compute the sums
of gray value change at each point in the analysis regions and
compare with that obtained from previous analysis. If at least
one of the sums is greater than a threshold, then the frame is
analyzed; else, it is skipped. Utilizing this criterion for
sampling, the system analyzes about 1/3 of the frames in a record
on the average and only 1/16 at the end of the contraction or
expansion period, when the heart wall shows very little movement.

3.6 Analysis of Consecutive Frames

Edge detection and boundary search in each subsequent frame
is guided by the model of the previously analyzed frame. The
analyzer measures an edge component parallel to the nearest boundary
at every point in the analysis regions. Then it searches each
analysis region for a boundary using the figure of merit given in
equation (1).

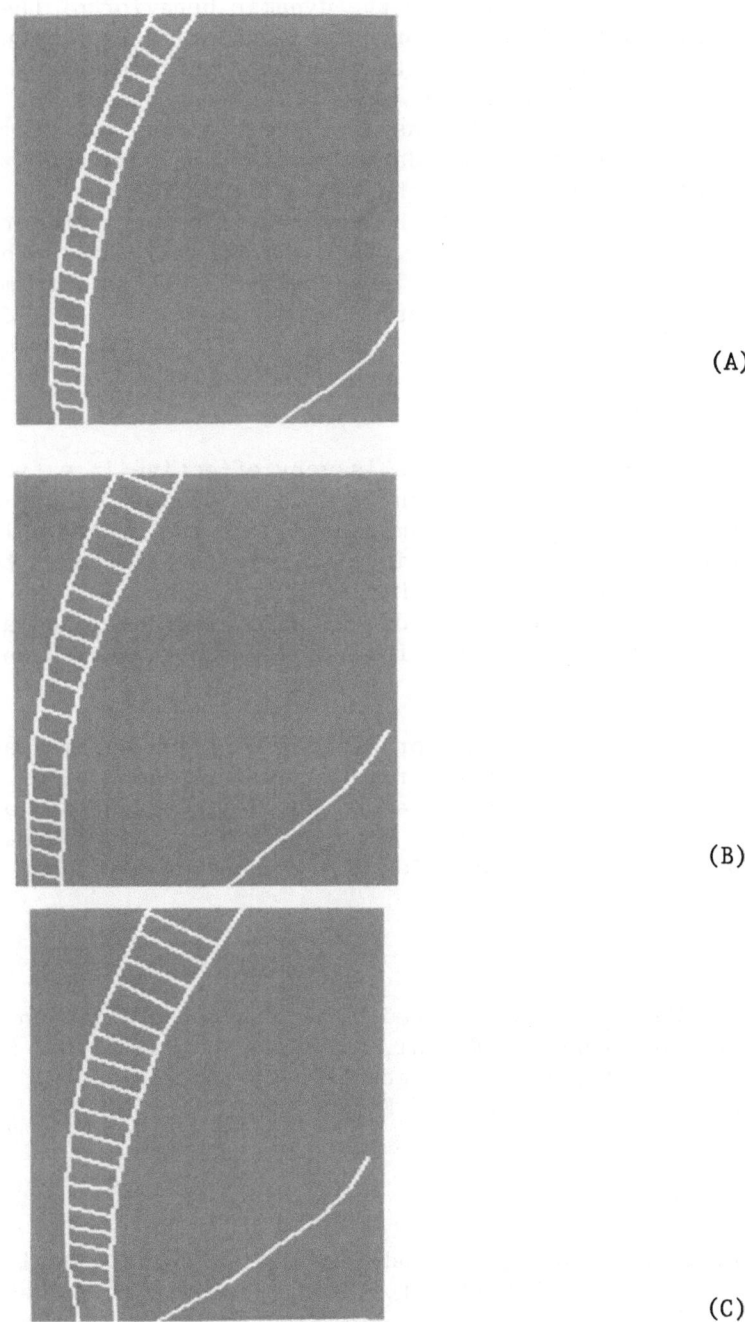

(A)

(B)

(C)

Fig. 7. Thickness variation and motion of a heart wall related to the (A) 10th, (B) 26th and (C) 34th frames with respect to the original frame.

3.7   Measurement of Wall Thickness

In order to measure the dynamic behavior of the heart wall,
we must determine the correspondence of each portion of the wall
from one analyzed frame to the next. At present, a human operator
specifies several small areas which seem useful for finding this
correspondence. Patterns in these areas in one frame are matched
to those along the boundary found in the next analyzed frame using a
a simple correlation technique. Figure 7 shows the correspondence
of several portions of the heart wall between the original frame
and the 10th (Figure 7A), 26th (Figure 7B), and 34th (Figure 7C)
subsequent frames. The line segments show us information about the
wall thickness variations.

4.   CONCLUSION

Plan-guided analysis is very effective in extracting significant boundary information from noisy dynamic images such as those
found in a cine-angiogram recording. The system which we have
developed is efficient since it selects (1) frames to be sampled,
(2) analysis regions within each frame, and (3) the directions of
boundaries to be detected. We have also developed a new and
efficient search method for finding smooth curves in a noisy picture.

Computation time for typical cine-angiograms was found to be
about 6 minutes for the first frame, 14 seconds for the next sampled
frame, and less than 1 second for deciding whether a subsequent
frame is to be skipped. We believe that this approach is essential
for the economic analysis of high resolution dynamic images.

5.   ACKNOWLEDGEMENTS

The research reported here is a cooperative work by members of
our laboratory, S. Kitani, M. Asada, M. Osada, and M. Ikeda. The
work was supported by a grant-in-aid for scientific research from
the Japanese Ministry of Education.

6.   REFERENCES

Ariki, Y., Kanade, T., and Sakai, T., "An Interactive Image Modeling
     and Tracing System for a Time-Varying Image Pattern,"
     TGPRL77-61, Institute of Electronics and Communications
     Engineers of Japan (1977) (in Japanese).

Chow, W. K., and Aggrawal, J. K., "Computer Analysis of Planar Curvilinear Moving Images," IEEE Trans. Comput. C-26:179-185 (1977).

Greaves, J. O., "The Software Structure for Reduction of Quantized Video Data of Moving Organisms," Proc. IEEE 63:1415-1425 (1975).

Jong, L. P., and Slager, C. J., "Automatic Detection of the Left Ventricular Outline in Angiographs Using Television Signal Processing Technique," IEEE Trans. Biomed. Engrg. BME-22: 230-237 (1975).

Kaneko, T., and Mancini, P., "Straight-Line Approximation for the Boundary of the Left Ventricular Chamber from a Cardiac Cineangiogram," IEEE Trans. Biomed. Engrg. BME-20:413-416 (1973).

Martelli, A., "An Application of Heuristic Search Methods to Edge and Contour Detection," Comm. ACM 19:73-83 (1976).

Montanari, U., "On the Optimal Detection of Curves in Noisy Pictures," Comm. ACM 14:335-345 (1971).

Yachida, M., Asada, M., and Tsuji, S., "Automatic Motion Analysis System of Moving Objects from the Records of Natural Processes," Proc. 4th Int. Joint Conf. Pattern Recogn., Kyoto (1978).

Yachida, M. et al., "An Interactive System for Digital Processing of Moving Images," Trans. of Institute of Electronics and Communications Engineers of Japan J61D (Oct. 1978) (in Japanese).

REAL-TIME IMAGE PROCESSING IN CT--CONVOLVER AND BACK PROJECTOR

H. Wani and H. Ishihara

Shimadzu Corp.

Kyoto, JAPAN

1.   INTRODUCTION

In recent years, Computed Tomography (CT) has caused a revolution in medical radiology. Many directions have been taken in reconstructing the tomographic image from the radiographic projections as reported by such authors as Bracewell and Riddle (1967), Ramachandran and Lakshminarayan (1971), and Shepp and Logan (1974). Most modern tomographic scanners use a method based on the so-called "convolution-back-projection" method. Generally, the amount of data collected during the radiographic scans is on the order of 300 points/projection and 256 projections, resulting in about 10^5 data points. These data should be convolved and back-projected during the scan. Therefore we have developed special processors for our CT scanner. In this chapter real time image processing in CT using a convolver and back-projector is discussed. An image display system for CT is also presented.

2.   IMAGE PROCESSING IN CT

Figure 1 shows a typical ray in the parallel-beam geometry for obtaining projection data. Let $X_{j,p}$ denote the projection data along the pth ray of the jth projection. The procedure of the convolution method is as follows:

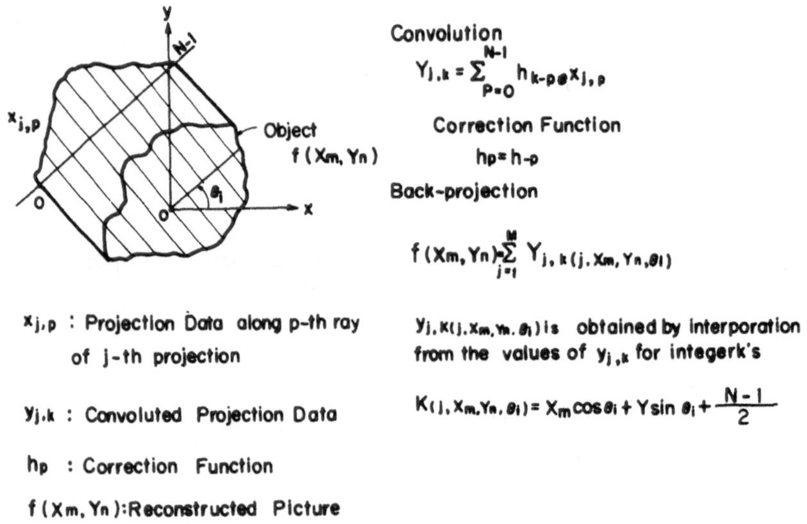

Fig. 1. Computed tomograph reconstruction steps for a typical ray using the convolution back-projection method.

(a) The calibrated projection data $(X_{j,0}, X_{j,1}, \cdots, X_{j,p}, \cdots, X_{j,N-1})$ for jth projection are convolved with the correction function
h_p (p = 0,1,...,N-1). Thus the data
$X_{j,p}$ (p = 0,1,...,N-1) for each projection is
processed by itself and is transformed in isolation from the rest of the projections into a
set of "convolved data" $(Y_{j,o}, Y_{j,1}, \cdots, Y_{j,N-1})$ where

$$Y_{j,k} = \sum_{p=0}^{N-1} h_{k-p} \cdot X_{j,p} \qquad (1)$$

(b) The density distribution at a given point
Xm,Yn in the object is obtained by back-
projection of the convolved data for each
projection as

$$F(Xm,Yn) = \sum_{j=1}^{M} Y_{j,k(j,Xm,Yn,\theta i)} \qquad (2)$$

3.   SYSTEM CONFIGURATION

A block diagram of our computerized CT scanner system is illustrated on Figure 2. The x-ray tube and detectors translate and rotate around the patient's head during a scan. The x-rays are divided into 16 pencil beams. After the x-ray photons pass through the object, they are detected by 16 detectors placed on each x-ray beam.

The logarithms of the detected signals are converted to digital signals by the A/D converter and then sent to the computer. At the end of each translation the raw data are calibrated. The calibrated data are sent from the computer to the convolver. The convolver performs the first stage of image reconstruction, working with one translation's data at a time. As the data are being con-volved, they are written back into the computer's memory.

For the final phase of image reconstruction, the convolved data are fed to the back-projector. When the back-projector com-pletes its task at the time that the final translation data are back-projected, the reconstructed image is stored in a cartridge type magnetic tape memory and is sent to the display unit through the computer.

4.   HIGH SPEED PROCESSOR

The steps of image reconstruction in our CT system are divided into two operations: (1) convolution and (2) back-projection.

4.1 Convolution

Convolution is performed in a special-purpose convolver which produces a sequence of dot products of vectors. The digital con-volution method is shown in Figure 3 and a block diagram of the convolver in Figure 4.

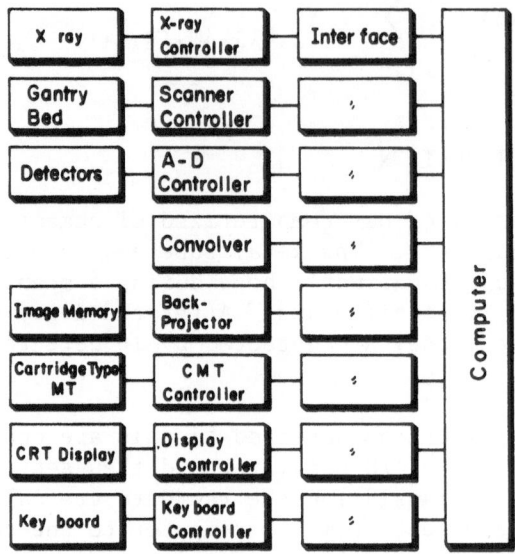

Fig. 2.   Block diagram of our CT computerized scanner system.

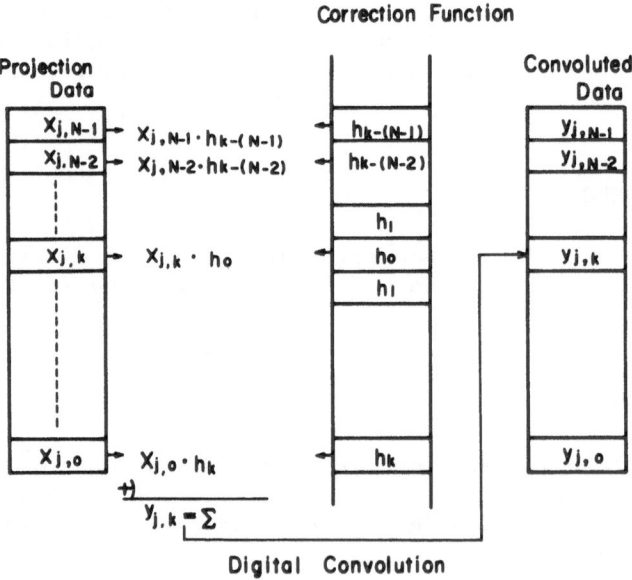

Fig. 3.   The digital convolution method.

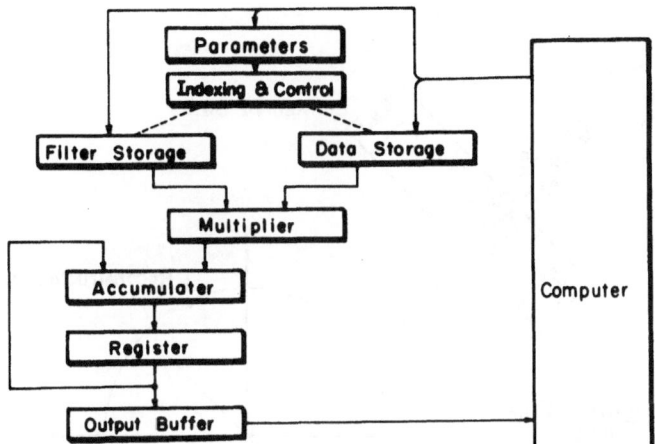

Fig. 4. Block diagram of the convolver.

As shown in Figure 4, the projection data to be filtered is stored in the data storage and the correction function is stored in the filter storage. Other information such as length of data, length of filter, number of output values, and input, filter, and output storage locations are also made available in the parameter storage between the computer and the convolver. Once the convolver is initiated, the convolver is under its own control which leaves the computer's central processor free for other operations.

The time for an accumulative multiply using this convolver is approximately 0.4 μsec. The maximum length of the data storage is 1024 points. The maximum length of the filter storage is also 1024 points.

4.2 Back Projection

To every point Xm,Yn in the reconstructed image plane there corresponds a value of $k(j, Xm, Yn, \theta i)$ as shown in Figure 1. Note that the value of $k_{integer}$ may not correspond to one of the values of k. However, the $Y_{j,k}$ is obtained by interpolation from the values of $Y_{j,k}$ for integer's.

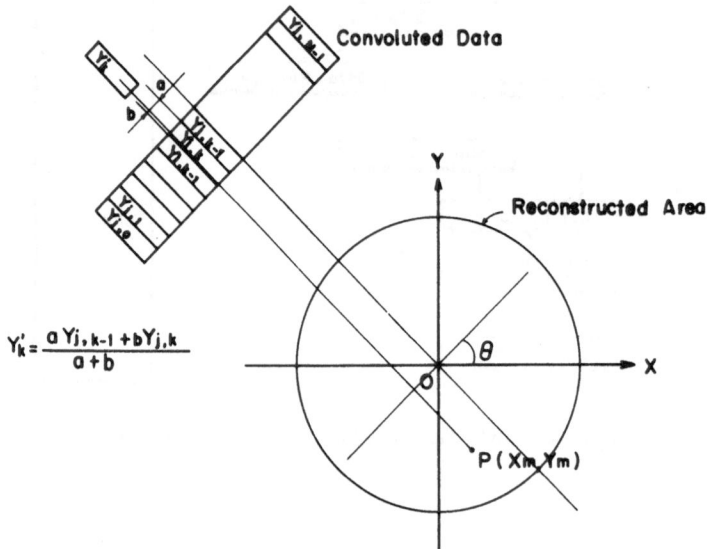

Fig. 5.   The back-projection method.

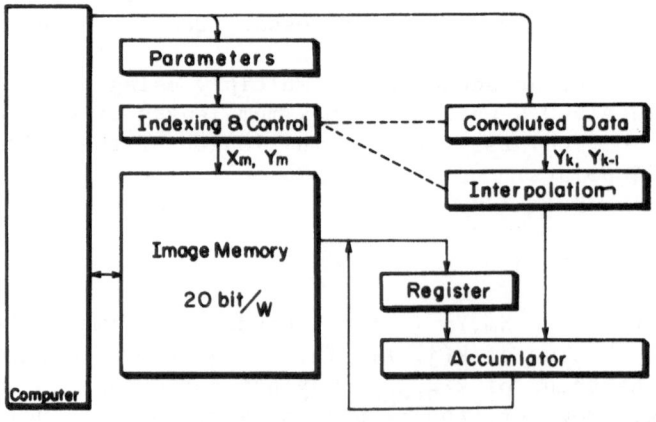

Fig. 6.   Block diagram of the back-projector.

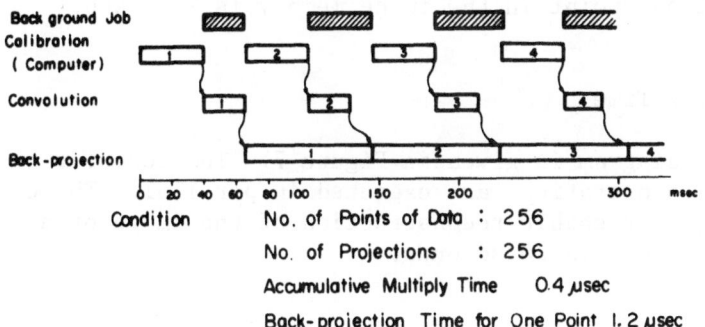

Fig. 7. Parallel processing of convolution and back projection.

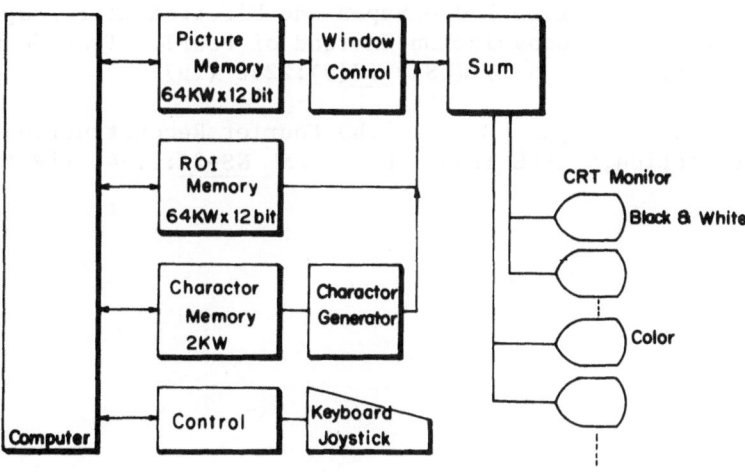

Fig. 8. Block diagram of the scanner display unit.

Figures 5 and 6 show the back-projection method and the block diagram of the back-projector, respectively. The convolved data are back-projected into the image memory which consists of 64K words using 20 bits per word. The time required for the back-projection of one point in the image memory is 1.2 μsec.

4.3 Processing Time

A timing diagram is given in Figure 7. The convolution and back-projection operations are executed in parallel. The convolver and back-projector enable reconstruction of the image of a section of a human head within 20 seconds.

5. DISPLAY SYSTEM

Figure 8 shows a block diagram for the display unit of our CT scanner system. The reconstructed image feeds to the picture memory in the display unit from the computer. The display unit has three types of memory: (1) picture, (2) ROI (Region of Interest), and (3) character.

6. REFERENCES

Bracewell, R. N., and Riddle, A. C., "Inversion of Fan-beam Scans in Radioastronomy," Astrophys. J. 150(2):427 (1967).

Ramachandran, G. N., and Lakshminarayan, A. V., "Three Dimensional Reconstruction from Radiographs and Electron Micrographs; Application of Convolution Instead of Fourier Transforms," Proc. Nat. Acad. Sci. (USA) 68(9):2236 (1971).

Shepp, L. A., and Logan, B. F., "The Fourier Reconstruction of a Head Section," IEEE Trans. Nucl. Sci. NS-21:21-42 (1974).

DETECTION OF THE SPHERICAL SIZE DISTRIBUTION OF HORMONE SECRETORY GRANULES FROM ELECTRON MICROGRAPHS

K. Baba, K. Miyamoto, and K. Kimura

School of Medicine, Dokkyo University, Tochigi, JAPAN

T. Kameya

National Cancer Center Research Institute, Tokyo, JAPAN

1.   INTRODUCTION

Optical micrographs are obtained from illuminating histological specimens cut into sections. Electron micrographs are obtained by the preparation of ultrathin sections and subsequent projection by the electron-optical illumination. Techniques developed in the field of stereology guide us to the way to statistically recover structural information from the pictorial information that is retained in the sectioned and/or projected figures.

For the purpose of high-speed, real-time processing, we must find a suitable information recovery rule for detecting the pictorial quantity from which structural information can be recovered when the shape of the objective structure is fixed but the size of the structure is variable. Three different rules for information recovery which apply to the analysis of the spherical size distribution of hormone secretory granules are described in this chapter. After the description, the method that is most suitable for high-speed processing is pointed out.

2.   MATERIALS AND METHODS

Specimens of medullary carcinoma of the thyroid were fixed in 2.5% glutaraldehyde and then in osmium tetraoxide, embedded in Epon 812, ultrathin sections prepared using a LKB Ultratome, and stained with uranyl acetate and lead citrate according to the mode

of Pease (1964). The three procedures for analysis that are used
in this study are described in detail below and summarized briefly
in Figure 1.

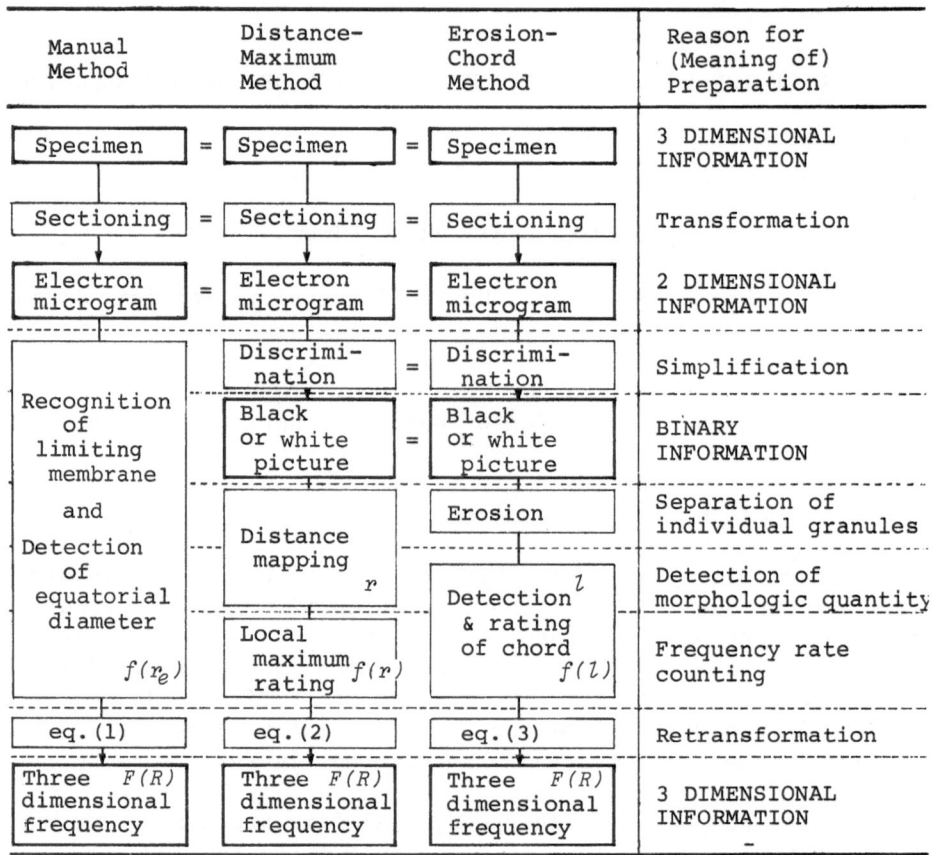

Manual Method	Distance-Maximum Method	Erosion-Chord Method	Reason for (Meaning of) Preparation
Specimen =	Specimen =	Specimen	3 DIMENSIONAL INFORMATION
Sectioning =	Sectioning =	Sectioning	Transformation
Electron microgram =	Electron microgram =	Electron microgram	2 DIMENSIONAL INFORMATION
Recognition of limiting membrane and Detection of equatorial diameter $f(r_e)$	Discrimination =	Discrimination	Simplification
	Black or white picture =	Black or white picture	BINARY INFORMATION
	Distance mapping r	Erosion	Separation of individual granules
		Detection l & rating of chord $f(l)$	Detection of morphologic quantity
	Local maximum rating $f(r)$		Frequency rate counting
eq. (1)	eq. (2)	eq. (3)	Retransformation
Three $F(R)$ dimensional frequency	Three $F(R)$ dimensional frequency	Three $F(R)$ dimensional frequency	3 DIMENSIONAL INFORMATION

Fig. 1. Flow charts of three analytical procedures.

2.1 Manual Method for Size Distribution Analysis

At least six good quality prints of electron micrographs taken
at a final magnification of 25,000 were selected for manual calibration of the hypothetical equatorial axis external to the limiting
membrane of the sliced granules. In order to eliminate the influ-
ence of the thickness of the ultrathin section on the apparent
frequency distribution of the size of the granules, the calibrated
values for each case were corrected by the following equation,

$$F(R) = \begin{cases} 0, & r_e < \dfrac{1}{2w}(\dfrac{t^2}{4} - w^2) \\[2em] \dfrac{f(r_e)}{2\sqrt{(r_e+w)^2 - r_e^2} - t}, & \dfrac{1}{2w}(\dfrac{t^2}{4} - w^2) < r_e < \dfrac{1}{2w}(t^2-w^2) \\[2em] \dfrac{2tf(r_e)}{\sqrt{(r_e+w)^2 - (r_e^2 - t^2)}}, & r_e > \dfrac{1}{2w}(t^2-w^2), \end{cases} \quad (1)$$

where F(R) is the distribution per unit volume of the spherical radius, R, of the solid granule; $f(r_e)$ is the distribution per unit area of the hypothetical equatorial radius, r_e, of the disk of the sectioned granule; w is the width of the gap between the margin of and the limiting membrane of the granule; t is the thickness of the ultrathin section. In this analysis, the thickness of the ultrathin section (t) and the width between the limiting membrane and the granule (w) were assumed as 50nm and 10nm, respectively.

2.2 Computer Method for Size Distribution Analysis

For computer analysis, at least six photographs, printed at a final magnification of 7,700, were used (an example is shown in Figure 2). The pictorial information in a 5.1x5.1cm photograph was transformed into a digital matrix composed of 512x512 picture elements and fed into a minicomputer (PDP-11/40) having 28K words of core memory using a mechanical picture reader (JCR-80/JEOL, Tokyo) and an analog-to-digital converter. This process trans- lates the pictorial information into digital information having 256 (8-bit) gray levels. The pixel spacing corresponds to 13nm in the original material. A 5-gray-level representation of the digitized pictures can be recorded on a Versatec dot-type line- printer (Figure 3). The configuration of the computer system used is shown in Figure 4.

2.3 Analysis

About nine hundred disks of the sectioned hormone granules were detected in each case, within a convenient area, by the fol- lowing two methods: (1) the distance-maximum method and (2) the erosion-chord method.

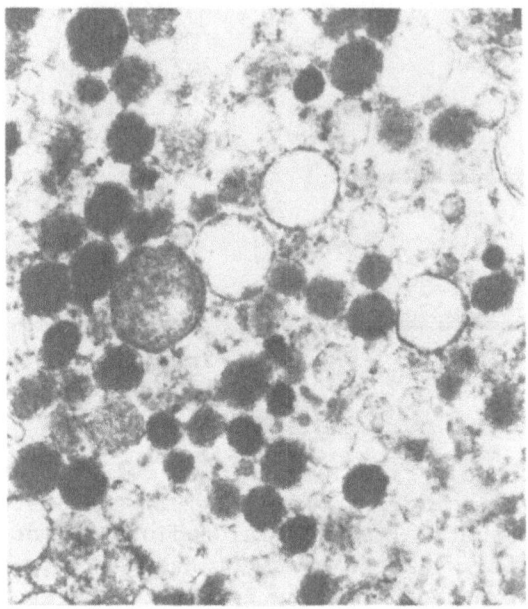

Fig. 2.   Typical electron micrograph showing hormone granules.

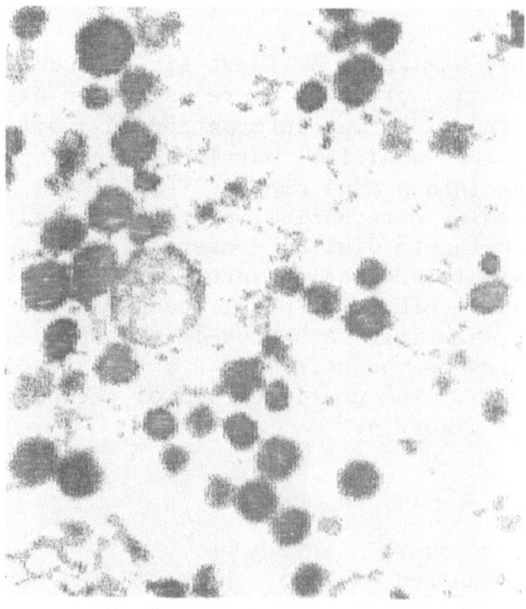

Fig. 3.   Digitized version of Figure 2.

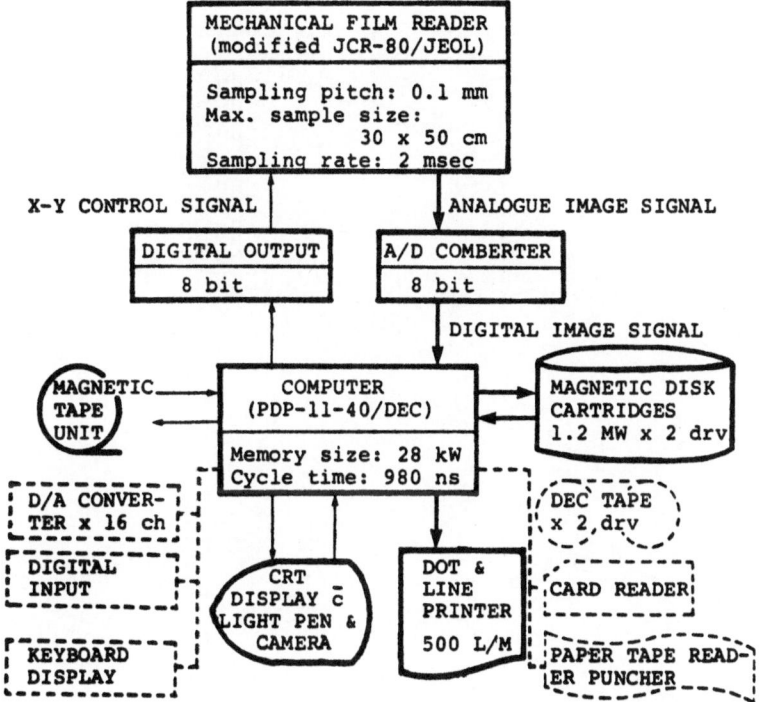

Fig. 4. Configuration of the computer system used.

2.3.1 *Distance-Maximum Method*

The digitized picture was transformed into a black and white picture (Figure 5) which was copied onto a magnetic disk for later use. The black and white picture was converted into a city block distance map (Figure 6) according to the method of Rosenfeld (1968). The pixel clusters composed of the local maxima in the city block distance domain were traced out by hand (Figure 6). To search the clusters of pixels which were evaluated locally as maxima, we designed an algorithm fundamentally based upon the grain-counting algorithm employed by a line association picture analyzer.

From the hypothetical frequency distribution of the values of the local maxima, i.e., tantamount to the radii of the disks representing the sectioned granules, the real frequency distribution, $F(R)$, of the spherical radius of the granules was stereologically computed by the following equation:

$$F(R_j) = \frac{1}{2\Delta r}[[\Omega_{ij}] + \frac{t}{\Delta r}[\delta_{ij}]]^{-1}f(r_i) \tag{2}$$

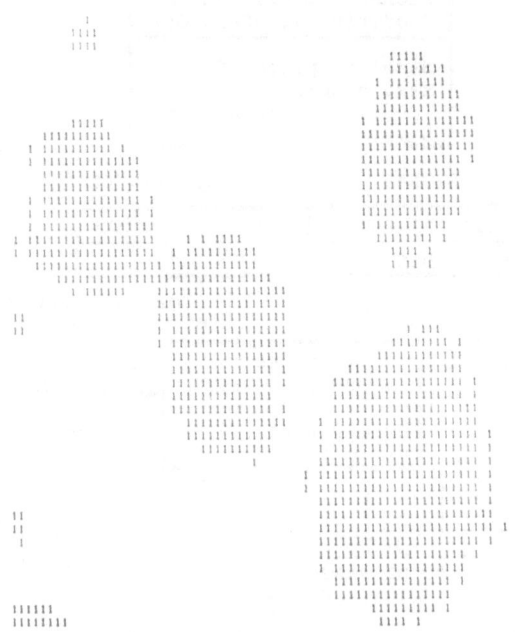

Fig. 5.   Black and white version of a portion of Figure 3.

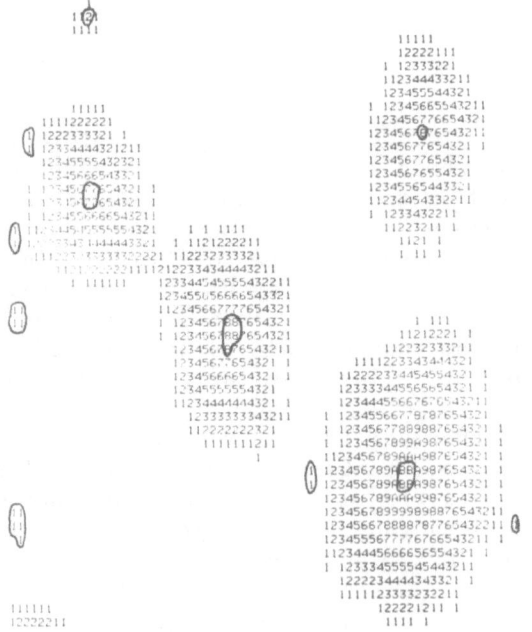

Fig. 6.   The local maxima of the city-block distance transform of Figure 5 are traced out.

where $F(R_j)$ is the spatial (three-dimensional) distribution function of the granules belonging to the jth class of the spherical radius; $f(r_i)$ is the two-dimensional distribution of the sectioned granules separated into the ith class according to the radius of their disks; Ω_{ij} is $\sqrt{j^2 - (i-1)^2} - \sqrt{j^2 - i^2}$; δ_{ij} is Kronecker's delta; Δr is a class interval of the radius of the sliced granule; t is the thickness of the ultrathin section. The quantity $[X]^{-1}$ denotes the inverse matrix of $[X]$. Equation (2) for t=0 is identical to the equation of Scheil (1935). In the present analysis, the thickness of the ultrathin section (t) was assumed to be 50nm. Further details on this approach are provided by Baba et al. (1973, 1976), Kameya et al. (1977), and Kimura et al. (1976, 1977, 1978).

2.3.2 *Erosion-Chord Method*

The transformed black and white pictures which had been stored in the magnetic disk were also processed by the erosion-chord method. The black and white pictures were eroded twice by a cruciform operator composed of 5 pixels of which the center was assigned as the characteristic point (Figure 7 insert). The eroded picture (Figure 7) was scanned and the lengths of the chords crossing the eroded disks of the sectioned granules were determined (Figure 8), increasing by one the variable that corresponds to the class of the chord length.

After correcting the distribution of the chord length for the shrinkage of the granule from using the erosion procedure, the distribution of the chord length was transformed into the spatial distribution of the spherical size of the granules using the following stereologic equation

$$F(R_j) = [\frac{\pi \Delta \ell^2}{4}[\Lambda_{ij}] + 2t\Delta\ell[\Omega_{ij}]]^{-1}f(\ell_i) \tag{3}$$

where, Λ_{ij} is (2i-1) provided that (i≤j); $\Delta\ell$ is a class interval of the chord length; $f(\ell_i)$ is the linear frequency of the chord length of the ith class. Here, Ω_{ij} is identical with that quantity given in equation (2). When the thickness of the ultrathin section is negligibly small compared to the radius of the granules, the rule for information recovery becomes very simple, because the remaining first term of equation (3) does not contain complex coefficients which would require a coefficient table. Under such conditions, t=0, equation (3) becomes the equation given by Spector (1950). In this analysis, we assumed a thickness of 50nm.

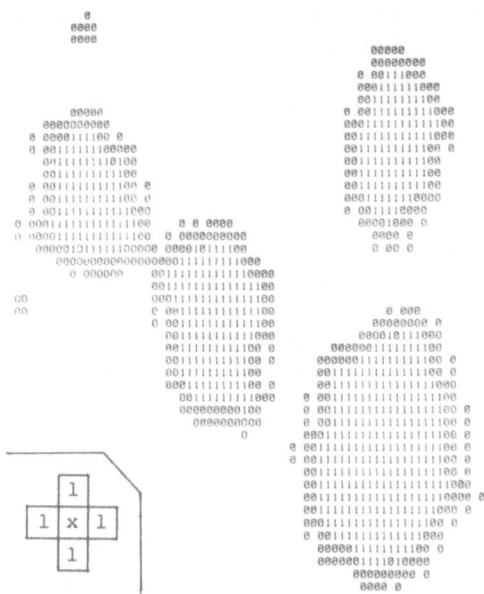

Fig. 7. Eroded version of Figure 5 where Ø indicates an eroded pixel. The insert shows the cruciform operator used where (x) is centered on the pixel which is being processed.

Fig. 8. Detected chords in Figure 5. The number associated with the last pixel of each chord is identical with the length of the chord crossing the eroded granule.

3.   RESULTS

Examples of the frequency distribution of the spherical radius of the granules determined by a manual procedure and the two computer procedures are shown in Figure 9. The curve delineated by the erosion-chord method shifts slightly to the larger side and shows a clearer shoulder at 200nm than that obtained by the distance-maximum method. Both curves obtained by the computer have a high peak of 150nm, with much lower peaks at approximately 200nm, 260nm and 310nm. The curves obtained by both computer methods are sharper than those manually generated.

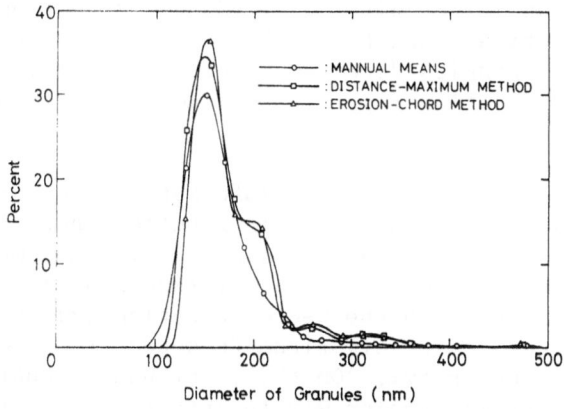

Fig. 9. An example of the distribution of the size of the spherical granule diameter computed using (1) manual means, (2) distance-maximum method, and (3) the erosion-chord method.

The time required for the determination of the hypothetical equatorial radii of the disks of the granules for each photograph was 2 to 3 hours by manual means. In addition, we spent more than 2 to 3 hours per case in the arrangement and the calculation of the raw data. The distance-maximum method required 50 minutes per photograph for processing including stereological information recovery and operation time. The erosion-chord method required less than 30 minutes per photograph to obtain the final result.

4.   DISCUSSION

Historically, pathologists have accumulated information on the correlation between hormone production and size distribution of hormone granules observed in electron micrographs. The size distribution has always been represented two dimensionally in terms of the radii of the disks of the granules but never three dimensionally.

The representation of the size distribution of the granules in
three dimensional terms may produce some initial confusion in comparison with historically accumulated knowledge. However, we propose to introduce the spherical representation of the size distribution, not only because of the ease in recovering stereological
information when using the computer, but also because of the ease
in detecting the hypothetical equatorial radii of the granules from
the electron micrograph in comparison to detecting the disks of the
granules *in toto* by manual detection. Moreover, the spherical
representation of the size distribution is definitely required when
the kinetics of hormone production is to be analyzed.

When the number of granules to be detected is small, however,
manual detection is not difficult. Manual detection becomes difficult and impractical when the number of granules is large. Thus,
we limited the granules detected based on the newly developed rule
represented by equation (1). This reduces time spent in detection
and in manual calculations.

Equation (2) makes possible information recovery appropriate
to the detection of the spatial distribution, both by classical
manual methods as well as by the distance-maximum method. The
distance-maximum method proves faster than manual methods. Since
we adhered for a time to the use of equation (2), our efforts to
find a simpler or faster algorithm to detect the radii of the
granules was unsuccessful. On the other hand, equation (3) led us
to establish a still simpler way to detect the desired information.
Further, the computer algorithm that evolves from equation (3) can
be extended to TV processing speeds.

The system that we developed offers some interactive correction
of the picture through the CRT display using a light pen and it may
be used for the interactive separation of the sliced granules. However, when we attempt automatic separation, we find that the separation procedure and the detection of the desired pictorial quantity
(the radius of the disk of the granule) can be simultaneously carried out by the distance mapping procedure of the distance-maximum
method. In the erosion-chord method, the separation procedure is
executed by the erosion step independent from the detection of the
pictorial quantity (the length of the chord crossing the sliced
granules) prior to chord length detection.

5.   CONCLUSION

The above discussion is a historical review of our analysis of
size distribution of hormone secretory granules. The erosion-chord
method which we have developed has not only been successfully
applied to high speed processing but also it may soon be realized
at TV rates in analyzing the required granule size distribution.

6.   REFERENCES

Baba, K., Amari, H., and Watanabe, Y., "Extraction of Ribosome
    Feature by Digital Filter," in Applied Electron-microscopy
    for Medical Science and Biology (Y. Yamada, et al., eds.),
    Tokyo, Ishiyaku-shuppan (1973), pp. 529-535 (in Japanese).

Baba, K., and Okayasu, T., "A Study on the Computerization of
    Quantitative Electron Microscopy: Application of the Distance
    Function to Ribosome Count," in Recent Progress in Electron
    Microscopy of Cell and Tissues (E. Yamada, et al., eds.),
    Tokyo, Igaku-shoin, Ltd. (1976), pp. 305-315.

Kameya, T., Shimosato, Y., Adachi, I., Abe, K., Kasai, N., Kimura, K.,
    and Baba, K., "Immunohistochemical and Ultrastructural Analysis
    of Medullary Carcinoma of the Thyroid in Relation to Hormone
    Production," Am. J. Pathol. 89(3):555-569 (1977).

Kimura, K., Hayashi, T., Miyamoto, K., Kanzaki, Y., and Baba, K.,
    "Morphometry (V): Stereology of the Sphere," The Cell 10(2):
    467-473 (1978).

Kimura, K., Kameya, T., and Baba, K., "Automatic Analysis with
    Computer Aid on the Frequency Distribution of Growth Hormone
    Granule Size," J. Electron Microsc. 25:194 (1976).

Kimura, K., Miyamoto, K., Kanzaki, Y., Ohgaki, M., and Baba, K.,
    "Morphometry (IV): Stereology of the Sphere," The Cell 9(8):
    283-294 (1977).

Pease, D. C., Histological Techniques for Electron Microscopy,
    (2nd ed.) Academic Press (1964).

Rosenfeld, A., and Pfaltz, J. L., "Distance Function on Digital
    Pictures," Pattern Recognition 1:33-61 (1968).

Scheil, E., "Statistiche Gufügeunfersuchungen, I," Z. Metallk.
    27:199 (1935).

Spector, A. G., "Distribution Analysis of Spherical Particles in
    Non-Transparent Structures," Zavod. Lab. 16(2):173 (1950).

THE ABBOTT LABORATORIES ADC-500^{T.M.}

J. E. Green

Abbott Laboratories

Dallas, Texas

1. INTRODUCTION

The ADC-500$^{T.M.}$ is a second generation differential white cell counter manufactured by Abbott Laboratories which performs a 500-cell differential on both normal and immature white blood cells (leukocytes). It also performs an assessment of red blood cell morphology and estimates platelet sufficiency at a throughput rate of 40 to 50 samples/hr (20,000 to 25,000 cells/hr) in unattended operation. The system consists of (a) a slide spinner for producing a monolayer of blood cells incorporating diffraction pattern sensing to adjust spin time for varying blood viscosities, (b) a stainer/loader which applies stain to the blood film under carefully controlled conditions and which inserts the stained sample slide into a small plastic holder, (c) an encoder which applies a human and instrument readable identification number to each holder and (d) a real-time analyzer which evaluates the sample.

Early research in the classification of blood leukocytes by image analysis used general purpose digital computers (Prewitt and Mendelsohn, 1966; Bacus and Gose, 1972; Young, 1972). Many seconds to a few minutes were required per cell. With the application of this research to the design of practical instruments for performing the leukocyte differential count, cell-processing time became a crucial parameter. The Corning Larc, an early automated differential counter described by Megla (1973) is a good example of this. In the Larc the pattern set by earlier research was followed: the image data were processed by a general purpose digital computer (in this case a PDP-8 minicomputer). An increase in cell analysis speed was achieved by careful software optimization

such that the cell analysis time was reduced to slightly less than
1 sec/cell with image data processing requiring between 0.50 and
0.66 sec/cell.

A departure from this trend was the Geometric Data Hematrak.
The Hematrak utilizes a hardwired special purpose computer which
performs all image analysis. The algorithms used in the Hematrak
are not based upon previous research results but utilize a set of
other image measurements which have been only partly described in
the literature (Miller, 1975; Preston, 1980). Also the Coulter
diff3 (based on the research of Ingram and Preston, 1970) utilizes
the iterative Golay transform for a portion of its analysis. This
transform is performed in a special purpose Golay processor attached
to a commercial minicomputer.

In the Abbott ADC-500 (Green, 1974) a cell analysis rate which
would permit a 500-cell differential to be performed in slightly
more than 1 min was desired. This analysis rate could be realized
by an extremely fast, very powerful but prohibitively expensive
general purpose computer. To reach a practical and economic solution, it was necessary that a substantial fraction of the image
data analysis be performed by a special purpose processor.

Two approaches have been suggested in designing special purpose image processing computers: (a) parallel processors and (b)
pipeline processors. The availability of low cost microprocessor
chips has made it economically feasible to apply array processing
methodology to image data. In this vein, Bartels and Wied (1976)
and Duff in a paper with Preston et al. (1979) among others have
proposed an array of microprocessors with one microprocessor for
each image element. While this approach would certainly yield an
extremely powerful and elegant solution, in designing an image
preprocessor for the ADC-500, it appeared that a sequential pipeline processor could potentially be equally fast, simpler and much
less expensive.

In studying the hardware/software design trade-offs, a
decision was reached that classification algorithms should remain
in software to permit easy update and improvement. With classifier development still progressing, it seemed likely that substantial improvements would be available from time to time. To
freeze the classifier in hardware would have precluded periodic
upgrades. For similar reasons the analysis of the density histograms to select thresholds used in scene segmentation remained in
software. The remainder of the scene analysis task, scene segmentation and feature extraction were considered to be stable enough
and suitable for incorporation into a hardwired preprocessor.

1.1 Rationale

For many years it has been traditional to perform the white
blood cell differential count at the 100 cell level. If one
attempts to estimate one component of the count such as the per-
centage of neutrophils by counting 100 cells, and the correct
value is 45%, sampling theory demonstrates that 95% of the time
the reported value will be between 35 and 55% (± two standard
deviations). As the first digit of the answer can change from
3 to 5, there are zero significant figures in this result. A
similar analysis yields the same conclusion for other cell per-
centages. (See Rumke, 1973.) To reduce the magnitude of this
problem, the ADC-500 counts a minimum of 500 leukocytes in each
differential. To maintain an acceptable sample throughput rate
for 500-cell differentials and to allow occasional 1000-cell differentials, a cell throughput rate of approximately 500/min is
necessary.

In order to allow reasonable stand-alone operation without the
necessity of reviewing every slide for cells classified as "other,"
we felt that the system should be able to classify every immature
leukocyte type routinely encountered in a large medical center's
hematology laboratory. With this ability, cells classified as
"other" or "unclassified" would truly represent those cells for
which no normal or immature category was appropriate. In order to
fully automate the differential, it was also necessary to provide
red cell morphology comments, platelet sufficiency evaluation,
automatic reading of sample identification from each slide and
automatic slide handling, oiling, etc. To provide consistent
accuracy, a semiautomated, controlled sample preparation system
was required. Thus the basic design parameters of the ADC-500
were established.

1.2 Rate-Limiting Factors

Cell throughput is limited by two major factors: (a) mechanical movement of the slide and damping of the resulting oscillations
to a tolerable level and (b) light available to the various sensors.
In order to minimize cell analysis time, several parts of the
system work in parallel: (a) the focus system, (b) the cell finder,
and (c) the multicolor high resolution scanner. This requires
that light passing through the objective be split into six separate
paths, three for the three color high resolution scanner, two for
the two autofocus sensors, and one for the cell finder array. In
addition, narrow bandwidth color filters are used to maximize
image contrast. This multiple splitting of light energy, added to
losses in the color filters, reduces the total energy available to
any single sensor. Compounding the problem, the insensitivity of

silicon photodiode arrays to short wavelength light requires sending
a large fraction of the total light energy to the 412 nm high resolution channel. As a result, a 150-watt xenon arc light source is
used. Considerable care was exercised in choosing beam splitters
so that minimum light is wasted. In several places dichroic beam
splitters are used to capture additional light energy which would
otherwise be absorbed in the narrow band color filters.

Let us return now to the problem of mechanical motion. Measurements of leukocyte density on spun blood films indicated that
the average leukocyte to leukocyte move would be 100-150 microns
with 98% of the moves less than 400 microns. Initial tests with
high performance stepping motors indicated that stepping rates well
over 2000/sec were easily obtainable. A step size of 8 microns was
considered to be adequate to center the cell in the 25x25 micron
high resolution field. For a number of reasons, every centered
scene will not contain an analyzable white cell. Occasionally, no
white cell is present in the low resolution field, so an arbitrary
move is made to an all red cell scene. Sometimes, double cells,
dirt, stain precipitate or other artifacts are centered. These
scenes must be discarded. At other times, the uncertainty in determining a white cell's location due to the 8 micron cell finder
pixel spacing results in the cell finder reporting cell coordinates
in error by 8 microns. When this occurs, the cell will be mis-
centered by approximately 8 microns and usually will overlap the
scene edge. Approximately 80% of all scenes will contain an analyzable white cell on slides from well prepared normal samples.
As low as 60% may contain analyzable cells on slides from samples
with extremely high white cell counts, clumped white cells, dirty
preparations, or the like.

When these discarded scenes are considered, to achieve a 500
cell/min analysis rate only 100 msec is available for the analysis
of each scene. Subtracting 34 msec for the two array scans (see
Section 3.5) for image acquisition leaves only 66 msec for both
mechanical movement and settling. Approximately 8 msec is required
for a 15 step average move (at 2000 steps/sec) and only 20 msec is
required for a 40 step maximum move. These numbers clearly pose
no problems for a 100 msec analysis cycle.

Tests of an early breadboard consisting of a Leitz Dialux
stand and a special x-y-z stage revealed that, for a single
8-micron step, vibrations with a peak-to-peak amplitude of 20-40
microns were excited. Ringing in excess of 0.5 micron continued
for up to 200 msec. It was evident that the step vibrationally
excited the structure; it was also evident that a conventional
microscope stand was poorly suited to high speed operation and a
new design was required. A design which overcomes these problems
will be described below.

2.   SAMPLE PREPARATION

Careful control over sample preparation is extremely important to the overall system accuracy of automated differential counters. Sample preparation requires (1) the production of a monolayer of blood and (2) the application of biochemical stains to this layer.

2.1 The Spinner

The use of a spinner to produce a monolayer distribution of blood cells on a glass slide was first reported by Ingram and Minter (1969) according to the original invention of Preston and Norgren (1971) who modified a spinner to accept glass slides which was originally designed for depositing photoresist material on silicon wafers in the microelectronics industry. Later, commercial versions for hematology use were produced by Platt, Perkin-Elmer, Corning, and Dyna-Tek.

There are several important factors in designing a spinner for hematology. Investigations by Ingram and Minter (1969), Staunton (1972), and Bacus (1974) have highlighted these. Foremost is the observation that, for increasing hematocrits and red cell counts, a longer spin time is necessary to produce the same cell separation. This relationship was extensively investigated by Bacus (1974) who demonstrated that it was primarily due to a viscosity phenomenon. Second, there is a spin-speed/spin-time trade-off, i.e., one can spin a smear at a higher spin speed for a shorter time (or vice versa) and achieve essentially the same cell separation. Third, it was discovered that the air flow in the spin chamber during and after spinning, although not adversely affecting cell separation, can have a profound effect on the resulting cell morphology. These effects were noted by Staunton (1972) and Bacus (1974) and have been further investigated by us. During spinning, the rate of acceleration and deceleration of the platen, the spin speed, platen streamlining and the configuration of the spin chamber interact to produce acceleration and air flow effects which can cause red cell pancaking, eccentric red cell central pallors, and an apparent increase in atypical lymphocytes and disrupted neutrophils. Spin chamber sealing to reduce air currents and reduce air exchange to a minimum have been found to be critical. In addition, spinning must be halted before the blood film begins to dry to allow the cells to reassume their natural shapes and air currents over the drying film must be eliminated to prevent distortions of leukocyte and erythrocyte morphology.

The Abbott spinner was designed with these factors in mind. A diffraction pattern sensor is used to detect the degree of cell

separation. Spinning is terminated when proper separation has
been achieved as described in Green (1978). In this way, varying
blood viscosities are accommodated automatically. In addition,
this device incorporates careful control of air flow to virtually
eliminate distortions of red cell and white cell morphology.

Considerable attention was also devoted to the safety of the
device when used with contaminated specimens. The spinner incorporates a positive air flow through the spin chamber and then through
a HEPA filter to capture and remove any aerosol which might be
formed during the spinning process. These features are combined
with an interlock which prevents access to the spin chamber for
5 sec to ensure that the air flow has purged the spin chamber
before the operator is exposed. The delay provides for 16 chamber
volume changes. In normal operation, the catch basin and platen
become coated with blood after a number of slides have been produced. A disposable catch basin and platen can be provided to
simplify decontamination. Finally, an electronic timer is incorporated which inactivates the spinner after 5 minutes of idle time
to prevent the HEPA filter from being prematurely clogged by room
dust.

2.2 The Stainer

The color of the nucleus and cytoplasm as well as the depth
of staining are key factors in both manual and instrument cell
identification. To repeatably produce cells with the same appearance, it is crucial that the formulation of the stain which is
applied to each slide be carefully controlled so that the staining
reaction will always proceed to the same end point. To achieve
uniform and consistent staining, the stain solution must be of uniform composition. Stain lots are monitored using high pressure
liquid chromatography to assure uniform composition. We have found
that the number of chromatigraphically distinct components in
Romanowsky stains exceeds that previously suspected. Limits have
been established on the concentration of key components and stain
lots are rejected or modified when these limits are exceeded.

The ADC-500 system uses a modified Ames Hematek stainer. In
this stainer fresh stain from a sealed container is applied to
each slide. The platen has been divided into two areas separated
by a drain trough. On the first area, alcohol stain solution is
applied to the blood film to fix the cells and begin the staining
process. The first station in this area is air predried ahead of
the advancing slide; when the slide has halted, stain rapidly
floods the gap. Rapid flooding is necessary to eliminate red cell
artifacts. When the slide leaves the first area, the first aliquot
of stain is peeled off and drains away. On the second area, a

stain buffer mixture is applied to the blood film to complete the staining process. A carefully controlled ratio of stain and buffer is premixed in a mixing coil as it is pumped into the capillary space between slide and platen. Premixing ensures uniform staining over the entire surface of the slide and a more uniform stain to buffer ratio.

The temperature of the entire platen is controlled at 19.5°C to assure uniformity of the staining reaction under different laboratory environments and to prevent red cell artifacts which occur at higher ambient temperatures. When staining is complete, the slide is lifted to an almost vertical position and is spray-rinsed by a series of nozzles. The slide then drains and dries in a vertical position. The vertical spray rinse and drain has been successful in eliminating stain precipitates. The vertical drying position allows additional drying time. This additional time in turn permits elimination of methanol from the rinse solution which results in a significant improvement in stain color.

Finally the slide is mounted in a molded plastic slide holder. The holder provides the medium on which a machine and human readable 6-digit identification is printed and permits slides to be stacked vertically for machine loading by the analyzer. The encoder which prints the code does not affect the slide quality or the quality of the results. It helps eliminate recording errors in the laboratory because it ensures that the proper sample identification is printed on the report form even if the order of the samples is inadvertently changed or an additional "stat" sample is inserted into a prearranged sample order. The encoder is also significant in that it allows batch processing of samples. Without a machine-readable sample identification, each sample would have to be fed one at a time into the analyzer with its corresponding hand-numbered report form or a complicated sample sequence would have to be rigidly maintained.

3.   REAL-TIME BLOOD CELL IMAGE ANALYSIS

The ADC-500 analyzer consists of (1) a computer-controlled microscope, (2) a special purpose image-processing computer called the "Preprocessor," and (3) a general purpose mini-computer (Nova-2 with 16K memory) as described by Green (1979A, 1979B). It is shown in Figure 1.

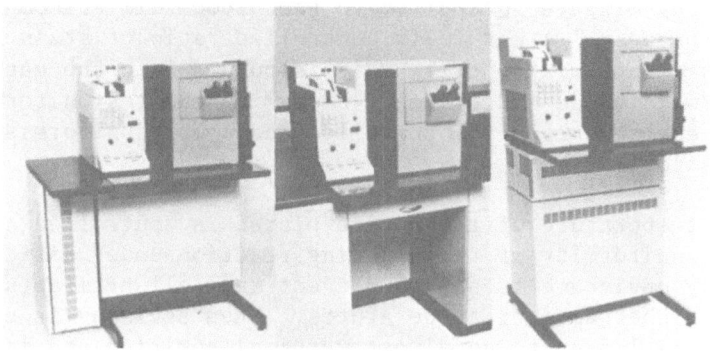

Fig. 1. The ADC-500 analyzer consists of three modules: the
power supply module (P); the computer module (C); the electro-
optics module (O), which can be rearranged to give the three
configurations shown.

3.1 The Computer-Controlled Microscope

This sub-system includes (1) input and output hoppers for
holder-mounted microscope slides, (2) x-y-z slide mover and automatic oiler, (3) oculars for viewing the blood film, and (4) a
printer for producing the laboratory report. The computer-
controlled x-y-z microscope stage is illustrated in Figure 2. The
inverted x-y stage is supported by three vibration isolation
mounts. The x and y moving elements are driven by stepper motors
at a normal rate of 2000 steps/sec, and a maximum rate of 8000
steps/sec. The stage steps in 8-micron increments and can settle
to ±0.25-micron movement (relative to the objective) in less than
35 msec. Mounted on the stationary stage member is a conventional
Leitz condensor and an objective lens carrier with a Leitz 40×,
1.0 NA oil plan-aprochromatic objective. The objective is driven
in the z axis by a linear induction motor and supported by two
crossed roller bearings. The y moving member of the stage incorporates a clamp for the slide/slide holder which indexes on the
top surface of the glass slide. This allows the objective-to-
blood film distance to be established within very close limits and
reduces the necessary objective movement to about 0.1mm.

Although containing most of the elements of a conventional
microscope, this assembly has a considerably different appearance
and form, in that the objective and slide are very stiffly
mechanically coupled. The mechanical path connecting the slide
and the objective in a conventional microscope is quite long and
disturbingly similar to a tuning form making it ideal for sustaining
vibrations. In contrast, the ADC-500 mechanical assembly has
reduced the coupling distance and increased the width to length

Fig. 2. The x-y-z stage is supported by three vibration isolation
mounts to prevent environmental vibrations from influencing analyzer
operation. Viscous damped stepping motors drive the x and y elements through ball-bearing nut assemblies. The assembly provides
a mount for a conventional microscope objective and condensor

ratio substantially, with all moveable components supported by
extensive bearing surfaces. The translational vibrations between
moving components excited by stepping motor input require some
friction for damping. This friction, if excessive, will reduce
the maximum slew rate because stepper motor torque decreases with
speed. Increasing friction also increases hysteresis; thus a compromise between settling time and speed-hysteresis is necessary.
Finally, a highly self-damping material (cast iron) was chosen and
the individual component parts stiffened with numerous webs to
reduce higher order vibrational modes in the structure itself.
The result is a structure with vibrational characteristics which
permits a 100 msec scene analysis cycle.

The remainder of the microscope consists of the optical and
optoelectronic elements shown in Figure 3. The light source is a
150-watt xenon arc lamp with integral parabolic reflector filtered
to eliminate both infrared and ultraviolet light. The light
passes through the condensor, the slide, the objective, and is
divided by two beam splitters into three components. The middle
component is directed to the focus sensors (two 64-element linear
photodiode arrays) after passing through a 525-nm bandpass filter.
The upper component is directed to the cell finder array after

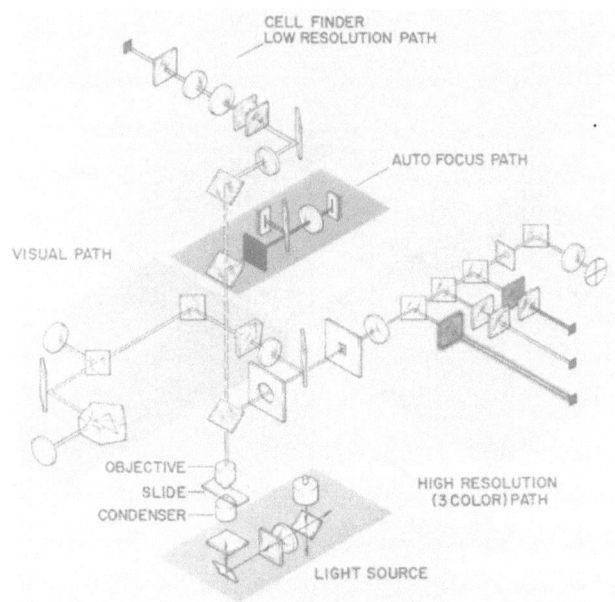

Fig. 3. The optical elements form six major chains: the light
source, the objective and condensor, the high resolution path, the
visual path, the auto-focus path and the cell finder path.

passing through a 560-nm bandpass filter. The lower component is
directed to the three primary sensors through bandpass filters at
412,525 and 560 nm, respectively. When the system pauses, as in
the review mode, a moveable mirror deflects the light from the
high resolution path into twin binoculars for manual observation.
Both the high resolution sensors and the cell finding sensor are
50 × 50 element self-scanned silicon photodiode arrays (Reticon
Corporation, Mountain View, Calif., USA). These photodiode arrays
have a number of advantages over traditional electron beam-scanned
image-sensing devices. For this application, the most important
of these are the device's geometric stability, inherent reliability
and lack of "lag." Geometric stability is especially important
because of the absolute requirement for superposition of the three
scanned high resolution images during scene segmentation. Long
term electronic stability is another important advantage of silicon
arrays. Although silicon devices exhibit drift with changing
temperature, once this drift has been characterized and compensated,
the device exhibits almost no drift with time. In contrast, vidi-
cons have a much smaller temperature drift, but exhibit highly
variable electronic characteristics as the phosphor ages. In
addition, an image can be burned into the phosphor of a vidicon
when used with high light levels. Silicon arrays do not exhibit
this problem.

The three elements of the optical path work independently. The focus system is a closed loop feedback circuit. The two 64-element focus sensors scan a strip near the high resolution field at 525 nm (green). At this color, red blood cells absorb strongly and provide a high contrast image. Each of the focus sensors is positioned slightly out of focus on either side of the image plane. A difference in image contrast between the two sensors drives the focus system into balance. If no contrast is detected, focus has been lost, and the system lapses into a search mode and scans through its entire focus range. In this case a computer interrupt suspends cell analysis until focus has been reacquired.

This focus method was selected because it could be implemented as an independent closed loop circuit. A more common autofocus method described by Kujoory et al. (1973) utilizes an iterative analysis of image sharpness (or high frequency image content) from the primary sensor while changing the focus in small steps. Using this latter method, focus position is adjusted until some focus figure of merit is optimized. Although considerably simpler in that it requires no additional sensors or circuitry, this method would require processing by the Nova 2 minicomputer and would be quite time-consuming. The dual out-of-focus sensor method is quite satisfactory if care is taken to assure that the two sensors scan exactly the same area. If the sensors are not carefully aligned, differences in the two scanned areas can cause the sensor outputs to balance at a position which is not in focus. Fortunately, the geometric stability of the silicon photodiode arrays makes alignment a stable adjustment.

3.2 Cell Acquisition

The cell acquisition detector scans a 400-micron square area just ahead of the high resolution field at 560 nm (see Figure 4). At this wavelength, only nuclear material absorbs strongly. Cells with nuclei, that is, leukocytes or nucleated red cells, are detected as dark spots on the array. The high resolution arrays scan a 25-μm square area (0.5 μm pixel spacing) at 3 wavelengths. During the high resolution scan, the cell acquisition detector is scanned and the x-y coordinates of all objects are reported by the preprocessor to the computer. A software buffer keeps track of these objects and selects the order of objects to be centered and analyzed.

Figure 5 illustrates the meandering path of the system as it searches for and acquires leukocytes. During the analysis of high resolution field A, the cell acquisition detector scans field A'. In A' object B is located and subsequently centered. Similarly, during the high resolution scan of object B, the cell acquisition

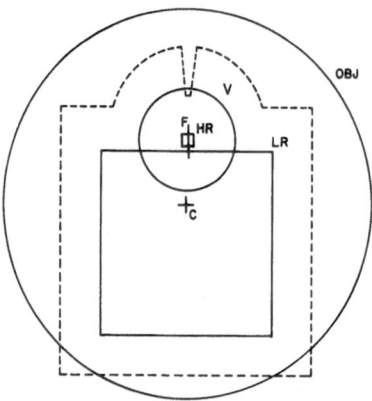

Fig. 4. The spatial relationships of the various fields are
illustrated in relation to the field of the objective (OBJ). The
high resolution (HR), the low resolution (LR), and the focus (F)
fields can be seen. The field visible through the binoculars, the
visual field (V), can be seen to be off-centered from the center
(C) of the objective field position of the substage field. The
dash line represents the position of the substage field stop.

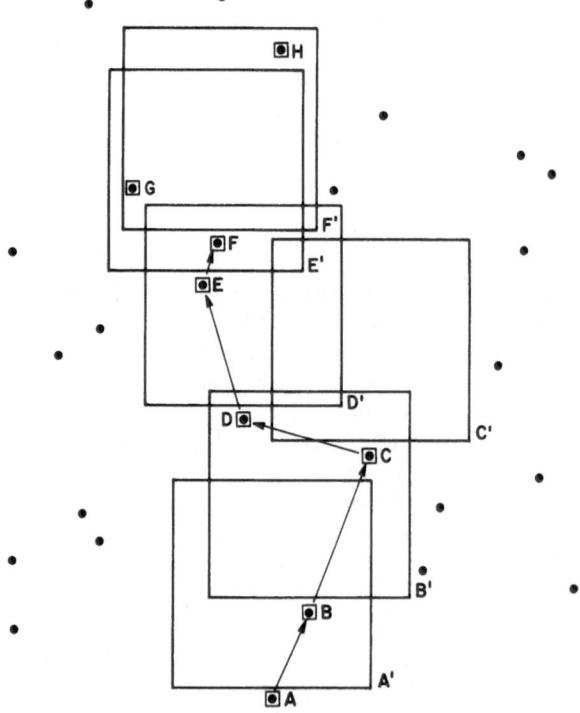

Fig. 5. The meandering path from cell to cell during cell acquisition is shown above. The text provides details.

detector scans area B' locating objects C and D. Notice that when
object C is centered and scanned there are no objects in area C';
however, the system remembers the location of object D and selects
it as the next object to be centered. If there are no objects
within the cell acquisition detector field and the acquired-object
software buffer is empty, then an arbitrary 400-micron Y move is
made, and the cell acquisition scan is repeated. With this
meandering track, it might be possible for the same cell to be
located on two successive passes across the slide. To prevent
this, the slide surface has been divided into corridors and the
minicomputer prevents corridor boundaries from being crossed.

3.3 High Resolution Data Analysis

The high resolution image data is first processed by the
Preprocessor which segments the high resolution image on a point-
by-point basis, groups together image elements from each object in
the scene, and extracts features from each object for later classification in the Nova computer. Figure 6 illustrates the data flow
in the Preprocessor. In the upper path, which we call the "histogram scan," histograms of the three images are compiled by the
Preprocessor and loaded into the memory of the computer. At the
same time, the image density data are buffered into the Preprocessor
image buffer. At the end of the histogram scan, the image data are
stored in the Preprocessor and the command to move to the next cell
is issued. Thus stage movement and settling can occur during cell
analysis and classification. The histogram scan requires 16.7
msecs. The three histograms are analyzed by a software program in
the Nova which determines the mean background density and four
thresholds used to segment the scene during the analysis scan.

The lower data route in Figure 6 is referred to as the
"analysis scan." The image data are sequentially unloaded from
the Preprocessor image buffer and analyzed by the Preprocessor
hardware. First the scene is segmented using the four thresholds
on a point-by-point basis, e.g., each point is determined to be
background, red cell, white cell cytoplasm or nucleus without
reference to the type of the surrounding points. A spatial filter
removes singular points which have resulted from errors in segmentation. As the scene is segmented, the Preprocessor logic groups
image elements which are part of the same object, assigning each
object encountered a unique "cell" number. At the same time,
features are extracted for each object using the assigned cell
numbers. Features which represent area, density, color and shape
of the various cellular components have been found to be the most
useful in cell classification (see Table 1). Table 2 shows a
partial list of additional features derived from the primary
features listed in Table 1.

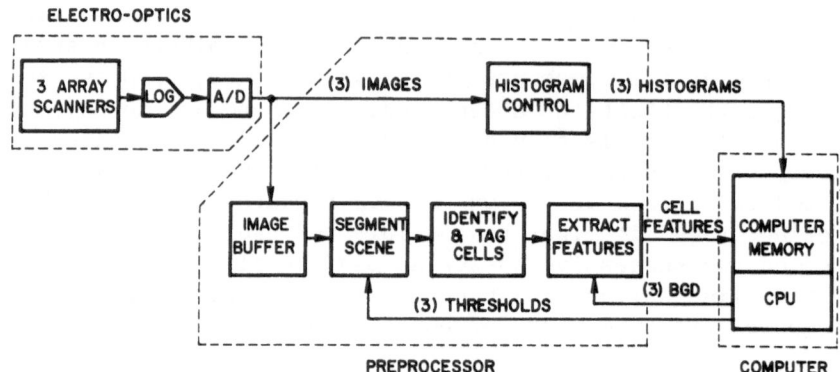

Fig. 6. Processing and analysis of the high resolution data is illustrated above. Images are acquired by the electro-optical system, scene segmentation and feature extraction are performed in the Preprocessor, and white cell classification and red cell and platelet analysis are performed in the computer.

Table 1 - Features Extracted by ADC-500 Preprocessor

White Blood Cells	Red Blood Cells
A_n nucleus area	A_c cell area
A_c cytoplasm	A_{cp} central pallor area
N_y nucleus density (yellow)	C_b cell density (blue)
N_g nucleus density (green)	C_y cell density (yellow)
C_y cytoplasm density (yellow)	CP_b central pallor density (blue)
C_g cytoplasm density (green)	CP_y central pallor density (yellow)
P_n nucleus perimeter	P_c cell perimeter
P_c cytoplasm perimeter	P_{cp} central pallor perimeter
	T target cell flag

Table 2 - Derived Features

$$A_c \quad \text{cell area} = A_n + A_c$$

$$D_n \quad \text{nucleus density} = N_g + N_y$$

$$D_c \quad \text{cytoplasm density} = C_g + C_y$$

$$C_n \quad \text{nucleus color} = N_g - N_y$$

$$C_c \quad \text{cytoplasm color} = C_g - C_y$$

$$S_n \quad \text{nucleus shape} = P_n^2/4\pi^2 A_n$$

$$S_c \quad \text{cell shape} = P_c^2/4\pi^2 A_c$$

3.3.1 *Scene Segmentation*

The scene segmentation method used is based on color algebra investigated by Green (1974, 1976B). It is a multicolored generalization of the traditional monochrome scene segmentation technique originally described by Prewitt and Mendelsohn (1966). Prewitt and Mendelsohn utilized the density differences between the nucleus, cytoplasm and scene background over a broad spectral range to separate these scene components. They found that if a density histogram of the scene was produced, the histogram contained peaks composed of the large number of image elements at about the same density from these three scene regions. Using the monochrome method, it is necessary that there be significant density differences between all regions of the scene one wishes to separate. Using our multicolor algebra it is possible to select several different wavelengths, each optimized to separate two or more scene regions which might not be separated using the monochrome method. For example, 412 nm separates erythrocytes from leukocyte cytoplasm, even though these two scene regions overlap at 525 and 560 nm. Complete scene segmentation is achieved by combining the information from the three wavelengths. Other related work in color algebra is given in Young and Paskowitz (1975).

In the Preprocessor each image point is assigned a category according to its color and density (the four categories are background, red cell, white cell cytoplasm or nucleus). The assumption is then made that image points of the same category which are juxtaposed are part of the same object. In order for this method to be successful it is necessary that (a) the different areas of

the image that must be segmented actually have a significant color
and/or density difference and (b) the wavelengths chosen to scan
the image reflect these color and/or density differences with sufficient signal-to-noise ratio. For peripheral blood cells stained
with Wrights stain, the first requirement is well satisfied. For
best results using our method, wavelengths must be selected which
maximize the contrast difference between these various areas. In
the ADC-500, the wavelengths selected are 412 nm, 525 nm and 560 nm.
At 412 nm hemoglobin in the erythrocytes absorb strongly while the
leukocytes are almost transparent. At 525 nm the eosin Y component
of the Wright stain absorbs maximally. The 560 nm wavelength was
selected as a compromise between the metachromatic staining of the
methylene blue components in Wright Stain and maximum nuclear/
cytoplasmic contrast. Histograms and scanned images at 560, 525
and 412 nm are shown in Figure 7 and in Color Plates 19 and 20.*

    Computer algorithms analyze the histograms and select thresh-
olds to separate the major scene components. First, the histograms are smoothed using a 1-2-1 weighted average to eliminate
noise. Depending on the number of peaks found in each histogram,
certain peaks are tentatively identified as background, red cell,
white cell cytoplasm, white cell nucleus or a combination of two
or more of these scene regions. The location of these peaks in
the scene density range and their relationship to other peaks in
the histogram are then tested to assure that the tentative idenfi-
fications are consistent with known patterns. If an inconsistency
is found, the tentative identifications are modified and retested.
In most cases, as in Figure 7, the identification of the peaks is
unambiguous. Finally, thresholds which separate the identified
peaks are set at or near the valley between the peaks. In ambig-
uous cases, the threshold location is tested to assure that it is
consistent with known patterns. The location of the background
peak is the most unambiguous feature of each histogram; its location is established first and is subsequently used as a reference
for the remaining peak and valley location tests.

    Nuclear isolation is the most difficult segmentation task to
accomplish and occasionally the correct nuclear boundary is not
established by either the yellow or the second green threshold.
This primarily occurs in some cells which have very dark blue cytoplasm or whose cytoplasm is densely packed with purple granules.
Examples of this are densely granulated promyelocytes, dark cytoplasm blasts and plasma cells, and dense basophils. The ADC-500
determines at least one threshold for each histogram and attempts
to set two in the 525 nm histogram. The 560 nm (yellow) threshold
is placed in or near the valley before the rightmost major peak;

---

*The color plates will be found following page 144.

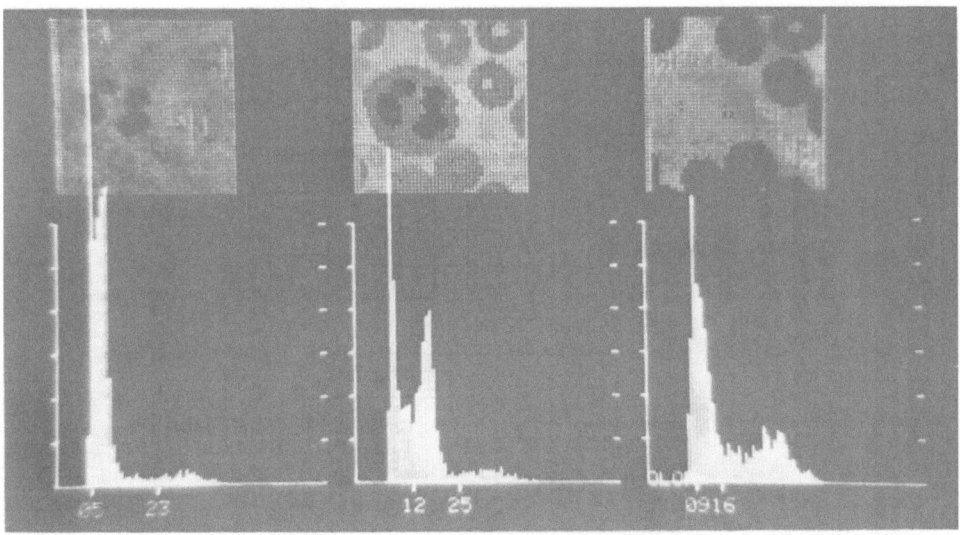

Fig. 7. Density histograms at 560, 525 and 412 nm (from left to
right) are shown along with cartoon displays of the corresponding
images. The calculated background density and threshold are
displayed for the 560 and 412 nm histograms (left and right).
The two thresholds are displayed for the 525 nm histogram (center).

in most images this threshold produces nuclear isolation. The
first 525 nm (green) threshold is placed in or near the valley
just to the right of the background peak; it effectively isolates
red cells and white cells from background. The second green
threshold is placed in the valley before the rightmost major peak
if the histogram has three peaks and such a valley exists; otherwise the second green threshold is set equal to the first. When
a second green threshold exists, in many cases it produces good
nuclear isolation where the yellow threshold has failed. The
412 nm (blue) threshold is placed in or near the valley between
the background-white cell peak (on the left) and the red cell peak
(on the right) so that good red cell isolation occurs but all white
cell areas are excluded. The effect of these thresholds on their
respective images can be seen in the cartoon image representations
in Figure 7.

3.3.2 *Spatial Feature Extraction*

The image data are processed serially in the Preprocessor as
it is read from the image buffer. In order to reassociate image
elements of the same type which were connected, one or more lines

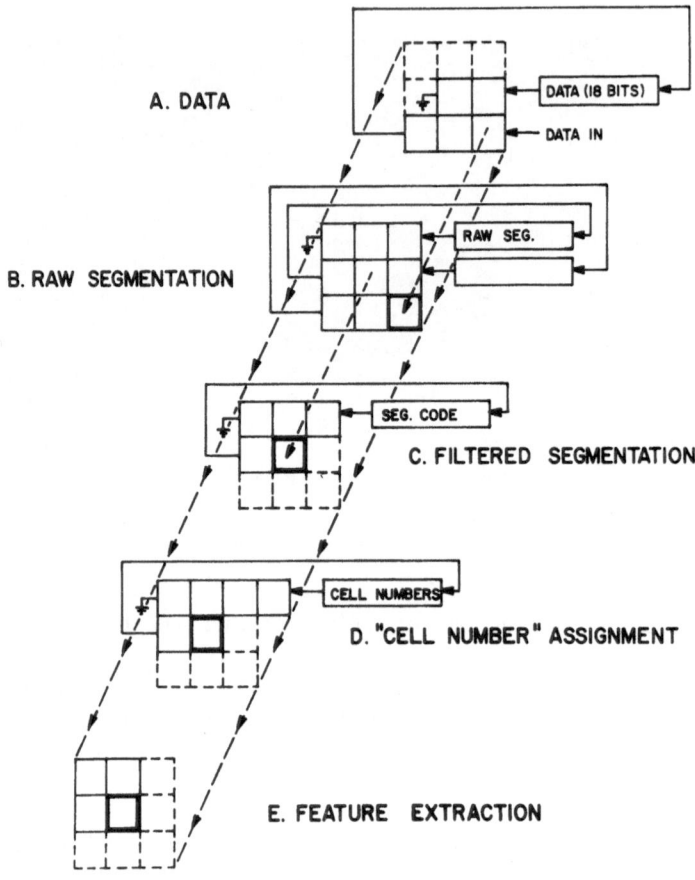

A. DATA

DATA (18 BITS)

DATA IN

B. RAW SEGMENTATION

RAW SEG.

SEG. CODE

C. FILTERED SEGMENTATION

CELL NUMBERS

D. "CELL NUMBER" ASSIGNMENT

E. FEATURE EXTRACTION

Fig. 8. The 3x3 analysis window is shown for Steps A through E of the scene analysis cycle.

of shift register delay are utilized to form a 3x3 element "analysis window." This analysis window is used throughout the remaining analysis and is illustrated in Figure 8 (steps A through E). At some points in the analysis the entire 3x3 window is not required; however, in Figure 8 the unused portions at a particular step are filled in to preserve continuity (indicated by ----). In Step D, an additional element was added to the 3x3 window to reduce the occurrence of multiply numbered cells. In Step A, the color algebra is applied to the data in the lower right-hand element of the analysis window. The data are delayed until they reach the center element in Step A. In B, the raw segmentation is also delayed until it fills the entire analysis window. The nine raw segmented elements in Step B are used to form the filtered segmentation of the center element in Step C which is delayed until it fills the remainder of the analysis window. Note that from this step onward (Steps C, D and E) the image elements before the center

element in the analysis window (that is, the bottom row and the
right hand element of the center row) are not needed and therefore
not present in the hardware. At Step D the Preprocessor groups
together image elements of the same type which are continguous.
Each object in the scene, either white cell or red cell, is
assigned a "cell" number in the order that it is encountered.
When the first image element in an object is encountered, it is
assigned a new cell number; when subsequent image elements in the
same object are encountered (by definition objects of the same
cell type which are continguous to the first element) they are
assigned the same cell number. Image elements above and to the
left of the center element of the analysis window are examined to
determine if they have the same object type as the center element.
If such an image element is found, the center element is assigned
the same "cell" number. As the image is processed by the Preprocessor, cell numbers will be transferred from element to element
from the first element in an object to the last.

     Because only a few points in any object are available to the
3x3 analysis window at any one time, objects which have U-shapes
or other irregular boundaries can cause multiple cell numbers to
be assigned to the same object. Likewise, in certain cells (such
as nucleated red blood cells), cell components can be juxtaposed
which normally do not occur. The Preprocessor recognizes these
situations, and issues what we call "equates" and "links," a series
of special flags to allow the software to correct these errors
later. Equates are used to combine the features for a particular
cell which were incorrectly compiled under multiple cell numbers.
When some cell border irregularity causes the cell number assignment logic to assign a second number to a cell, this number is
propagated until two adjacent image elements in the same object
are detected which have different numbers. This condition is
recognized as an error, and the first number is changed so that
the remaining part of the cell is assigned only the second number.
However, a part of the cell's features were compiled under the
first cell number. The equate instructs the computer to add the
features stored under the two cell numbers before further analysis.
The equate function is also used to flag a cell touching the scene
edge, by definition cell number zero. Cells which have an equate
to zero are excluded from further analysis. In a similar fashion,
links are used to flag the detection of nuclear material (by definition "white cell" material) and central pallor in red blood cells.
White cells and red cells have a separate progression of cell
numbers assigned. In the case of a nucleated red cell it is
necessary to associate information about these two major cell types.
The link is used to link features from a given red cell to features
from a given white cell. The linked features are later tested to
detect the presence of a nucleated red cell. In a similar fashion,
links are used to associate features from a red cell's central
pallor with the appropriate red cell.

The assigned cell numbers are not stored or saved for the entire image. One line of shift registers is used to delay the numbers only long enough to allow the cell numbers for the following row of the image to be assigned. These numbers are then discarded. As the cell number is assigned to each image element, features for that object are compiled. In the Preprocessor there is an N-word buffer for each cell number corresponding to the N features derived from each object. When the image element in the center of the analysis window is assigned cell number M, the feature buffer in the Preprocessor for cell number M is updated using the data from the image element. For example, the word in buffer M containing cell area is simply incremented by one; the three words containing integrated density are incremented by the amount of the three density values, etc. At the end of the analysis scan (which in fact does not involve a scan of the arrays) the features for all objects in the scene and the equate and link list are loaded from the Preprocessor feature buffer into the mini-computer memory.

Different features are collected for erythrocytes and leukocytes. Central pallor area is measured for each red cell. Because central pallor elements are indistinguishable from background elements, all elements which receive the "background" segmentation tag are also assigned "cell" numbers. At the beginning of the scene, the scene background is assigned background cell number 0; subsequent red cell central pallors which are not contiguous with the scene background are assigned numbers 1, 2, etc. Central pallor background points are then associated with corresponding red cells using an appropriate link.

3.4 Cell Classification

After the Preprocessor has completed its analysis of the image data, the Nova 2 minicomputer in the ADC-500 assumes the major role in cell analysis. A software program resolves the links and equates, removes any object whose border touches the scene edge, compiles statistical data on platelets and red cells, groups features together for the main white cell (or cells) and classifies the white cell. The program uses a hierarchically structured tree classifier which successively narrows the classification of each cell until a final classification is reached. The tree structure was derived by Cain and Anderson (1977) using the method of and computer programs supplied by Prewitt (1972). As stated by Cain, "structural articulation proceeds *de novo* using similarity or proximity in feature space to produce a taxonomy responsive to structure detected in the raw data. Two class discriminant functions are derived at each node in the structure using the classical multivariate statistical approach (LDA). Stepwise selection of

features at each node according to their linear discriminatory
power accompanies parametric calculation (by maximum likelihood
rule) of the two class discriminant function." Twenty-four cell
classes are identified and are collected into 11 groups for
reporting purposes. When the classification is complete, features
for the object are checked against feature limits for the class
selected. Objects not falling within these limits are discarded
as trash. The hierarchically structured classifier was selected
because we found that overall accuracies were improved 5 to 10%
over those provided by a 24-way LDA classifier.

The red cell features listed in Tables 1 and 2 are extracted
from each scene in parallel with the white cell features. Individual red cells are not classified; instead distributions of red
cell parameters representing size, shape, hemoglobin content and
polychromasia (blueness) are constructed for the entire sample.
The size distribution is called a Price-Jones curve. When the
analysis of 500 leukocytes is complete, the red cell distributions
are analyzed. The smear is judged to be normal or "morphology
comments" of 0 to 3+ are printed on the report form for micro,
macro, aniso, poik, hypo, and poly. Platelets are counted as white
cell and red cell data are analyzed. At the end of red cell evaluation, the average number of platelets detected per scene is compared with a normal range. Values outside this range are printed
as increased or decreased platelets on the report form.

3.5 System Timing

Within the high resolution system several tasks also proceed
in parallel. The 100 msec scene analysis cycle is divided into
six 16.7 msec array scans. The beginning of stage movement
corresponds to the beginning of the system cycle. The steps are
(1) new cell finder scan, (2) stage movement, (3) stage settling,
(4) image wash, (5) image scan, (6) histogram threshold calculation, (7) Preprocessor analysis scan, and (8) cell classification.
At the beginning of a cycle, the computer cell buffer contains the
location of the next white cell to be analyzed (produced by
previous low resolution array scans) and the command to move to
that location is issued. The move usually requires 5-20 msec. A
40 msec settle time follows. One scan is then used to "wash" the
blurred image off of the photodiode arrays. The last frame of the
cycle is a Preprocessor histogram scan. During this scan, (a) the
histogram is formed by the Preprocessor, (b) the image data are
read into the Preprocessor image buffer and (c) the cell finder
(low resolution) array is interrogated to locate subsequent white
cells to be analyzed. At this time the move-settle-wash-scan
cycle begins again; however, in parallel the computer is analyzing
the information just stored. The histogram thresholds are calculated, a Preprocessor "analysis scan" is initiated, and the

classification programs proceed. By the time another histogram
scan occurs, the data from the previous scan have been analyzed.
One can see that during the histogram scan three cells are actually
being processed by the analyzer; the first cell is being classified
by the software classifier, the second cell is being scanned by the
high resolution data system and the third cell is being located by
the cell acquisition system.

      This technique is usually called pipeline processing. The
analysis of each scene actually requires more than 200 msec, but
there is more than one scene in the analysis pipeline at any one
time. Thus, the average throughput rate of the pipeline is 100
msec/scene. Sometime during this cycle other system functions are
coordinated by the computer. For example, the cell count, various
system faults, interrogation of the control console and the like
occur. Between slides other functions such as slide unloading and
loading, slide oiling and report form printout are also coordinated.
In addition, the minicomputer coordinates both the review mode and
the calibration/test mode of the instrument.

3.6 Review

      At the end of a run comprising many slides laboratory personnel
will usually decide to review certain samples which have a non-
normal printout. This review can be performed using an ordinary
microscope. It can also be performed using the review mode of the
ADC-500. In the review mode, any suspect slides are reloaded into
the analyzer. The operator selects the types of cells he wishes
to review by pressing the appropriate cell type buttons on the
control panel. The instrument then locates and classifies cells
as in the automatic mode, except when a selected cell type is
encountered the instrument pauses and presents the cell for
operator review. At the end of a review, a new 500-cell differential has been performed and a new report form printed. When a
cell is presented and the operator disagrees with the classification, he may change the classification by pressing the appropriate
cell class button. Thus the new differential has been verified and
edited by the reviewer. The review may be terminated at any time.
When this occurs, the results to that point are printed.

      The calibration/test mode also deserves mention. The ADC-500
is sufficiently complex that simple manual tests are inadequate to
determine whether it is in calibration and functioning properly.
Therefore, we have made use of a series of resident software test
programs working in conjunction with a vacuum-deposited aluminum
test slide which contains a number of test targets appropriate for
each test program. These programs check thelight intensity, the
density calibration of the three high resolution sensors, focus

operation, stage accuracy and settling, cell finding and the
various instrument accessories. Further detailed tests of the
electro-optics, preprocessor and computer may be performed using
auxillary program storage.

4.   RESULTS

     Table 3 presents a confusion matrix comparing the ADC-500 cell
classification with that of a human observer. In many cases where
the system and the human disagree, there is only a small difference
in cell maturity between the two classes selected. The data for
this test were a carefully randomized sample of human identified
cells (selected on an as-encountered basis) with equal numbers of
cells from each class, the members of which were obtained from a
large number of slides. In this way we attempted to estimate the
accuracies from a much larger number of clinical samples. Even so,
in some cases confusions shown have not proven to be typical of
system performance.

     Table 4 illustrates the result of a study of system precision.
Five hundred cell differentials were performed on 86 slides prepared from the same blood sample. Mean cell fraction, coefficient
of variation (CV) and theoretical minimum coefficient of variation
predicted from sampling theory are tabulated for 500 and 100 cell
differentials. The coefficient of variation is CV = $S/P \times 100$
where the standard deviation S is given by

$$S = \left(\frac{P(1 - P)}{N} \right)^{\frac{1}{2}}$$

and P is the true fraction of the subpopulation in the total population, and N is the number of members of the total population
examined. This is the minimum S and CV that can be expected with
sample size N if all other errors are zero. Classification errors
tend to increase the measured CV. As can be seen from Table 4,
except in the case of band neutrophils, the standard deviation and
coefficient of variation agree very well with the minimum predicted
by sampling theory for 500 cells. Note that for a 100-cell differential, the theoretical minimum coefficient of variation is 2.2
times that for a 500-cell differential. Thus it appears that the
increase in precision predicted by increasing the cell count from
a 100 to 500 is actually realized. The larger observed CV tabulated for eosinophils and basophils results from the reported
results being rounded to the nearest percent.

Table 3 - Human Versus ADC-500 Cell Classifications

Human Call	ADC-500 Call														TOTAL	% CORRECT
	SEG	BAND	META	MYELO	PROMY	BASO	EO	MONO	PRO MO	BLAST	LYM	PLASMA	PLA. ATY.	ATY. LYM.		
SEG	264	39				1	4		1			1		1	311	85
BAND	23	98	6			2	4	2						1	137	72
META	1	12	95	2	1	4			1		2			4	122	78
MYELO		1	10	101	4		5	6	1	1				3	132	77
PROMY			3	9	108	5			1	20		4	4		154	70
BASO				1	1	144				4	8	1	4	3	166	87
EO	1			2		3	147				1				154	95
MONO			1			3		168	4	10	1			16	203	83
PRO MO								7	105	5	1			6	124	85
BLAST					10	10		2	8	170	3		24	16	243	70
SL						8					262			1		
ML						3					137			1	528	95
LL						4		1		1	103		2	7		
PLASMA					1	6					2	85	27	2	123	69
PLA. ATY.					5	12			1	11	1	6	93	3	132	70
ATY. LYM.					2	16		18	2	28	24	6	18	251	365	69
Total	289	150	119	115	133	218	160	204	124	240	543	103	171	315	2,884	

Table 4 - Precision of 500-Cell Differentials[a]

Cell Class	Mean	Coefficient of Variation		
		Observed	Theoretical for 500 cell	Theoretical for 100 cell
SEGS	48	4.4	4.6	10.5
BANDS	4	36	22	50
LYMPHS	36	0.7	6.1	13.3
MONOS	8.6	15.2	15	33
EOS	2.7	34	28	60
BASOS	0.7	84	56	120

[a]Eighty-six slides were prepared from the same blood sample and 500-cell instrument differentials performed. The observed coefficient of variation and the theoretical minimum coefficient of variation for 500 and 100 cell differentials are tabulated for the six normal cell types.

5.   SUMMARY AND CONCLUSION

The ADC-500 is a new blood cell differential classifier. It performs 500-cell leukocyte differentials on both normal and abnormal cells, evaluates red cell morphology and estimates platelet sufficiency at a rate of 40 to 50 samples per hour in stand-alone operation. The ADC-500 system consists of a spinner which prepares a uniform blood monolayer on a slide, a stainer which reproducibly stains the slide with Wright's stain, an encoder which attaches an instrument and human readable identification to the slide and an analyzer which accepts a stack of up to 50 slides, evaluates these slides and prints the results and the slide identification on report forms. The system's analysis rate, which represents a 5- to 10-fold increase over other commercially available differential counters, requires a number of specialized techniques for its realization.

One key to real-time, high-speed performance is the development of a special x-y slide positioning stage which can move to a new cell and settle in 50 msec. Another is the high degree of parallelism used in the system structure and the pipelining of the data processing. A third is the development of uniform and repeatable sample preparation modules. Within the analyzer

module, the autofocus, white cell acquisition and high resolution
cell analysis systems are independent and operate in parallel. At
the same time within the high resolution cell analysis system, one
cell is acquired; the digitized image of a second processed; and a
third is classified using pattern recognition techniques. All of
these tasks, except focus, are under the control of a minicomputer
system. Tests of the system reveal good accuracy and an improvement in precision due to the increase in the number of counted cells.

The sequential, pipeline Preprocessor segments the three color
image data from the ADC-500 optics one image element at a time,
groups together image elements from each object in the scene and
extracts features from each object. The processing occurs at television frame rates, requiring 16.7 msec to process the entire
image. This speed was instrumental in allowing the ADC-500 automated differential analyzer to perform routine 500-cell differentials. The Preprocessor also contains hardware which simplifies
compilation of the three color histograms. The segmentation algorithms implemented in the preprocessor are multicolor extensions
of the classical monochrome density histogram threshold method.
For most cell image analysis tasks, a sequential pipeline processor
of this type should be more economical and as fast or faster than
a parallel processor. In the Preprocessor, the features which are
compiled are relatively simple. The classical features of size,
shape, density and color of the nucleus and cytoplasm have been
found to be powerful discriminants for leukocytes. Using the same
image-element by image-element method, it is possible to compile
more complex features. With the exception of iterative image
operations, such as Golay transforms, we have not encountered any
cell features which could not be implemented relatively simply in
the Preprocessor.

In developing classification algorithms for an automated differential instrument, it is important to note that there is no
absolute standard of "truth" which can be used. Normally, the
classification of an "expert" human observer is defined as truth.
Unfortunately, on a distressingly large number of occasions, two
qualified human observers will not report the same classification
when viewing the same cell. Bacus et al. (1973) extensively
investigated this classification problem for the six mature leukocyte types and found there was significant observer disagreement.
When the choices available to the observer is increased from the
6 mature cell types to 20 or more mature and immature types, the
frequency of disagreement increases substantially. Confusions in
cell identification by the human expert tend to degrade the
training process by presenting the classification programs with
misleading data and make the tested accuracies appear lower. To
our knowledge, there is no satisfactory solution to this dilemma.

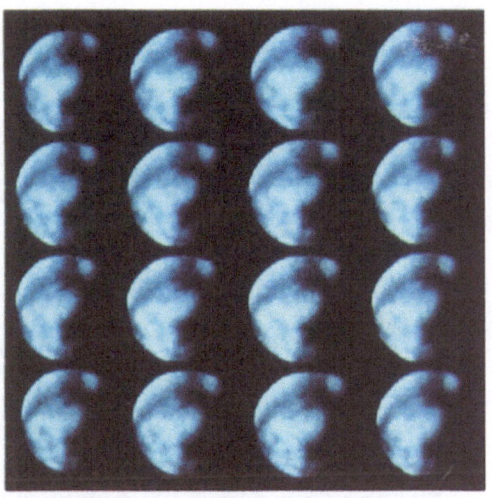

Plate 1
Nuclear ventriculograms
(Sternberg)

Plate 2
Contour enhancement
(Sternberg)

Plate 3
Outline smoothing
(Sternberg)

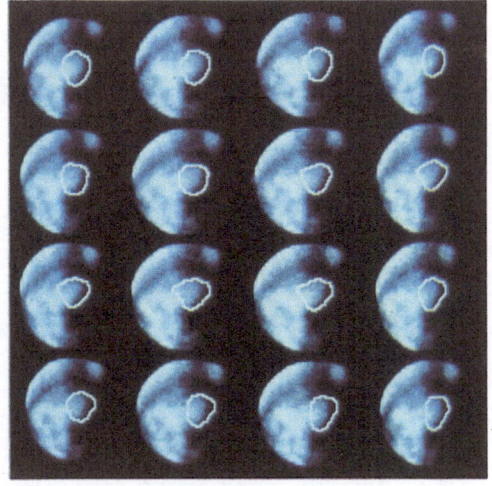

Plate 4
Left ventricle located
(Sternberg)

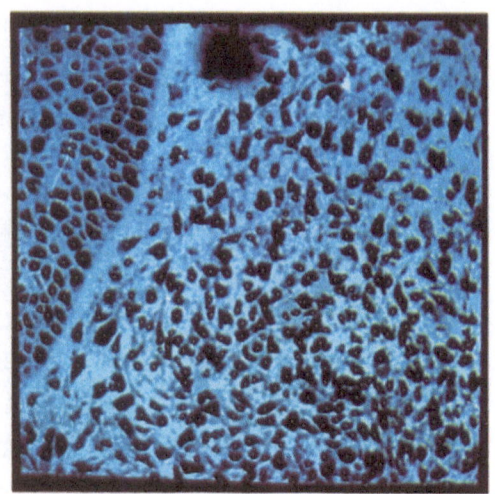

Plate 5
Newt limb-bud tissue
(Sternberg)

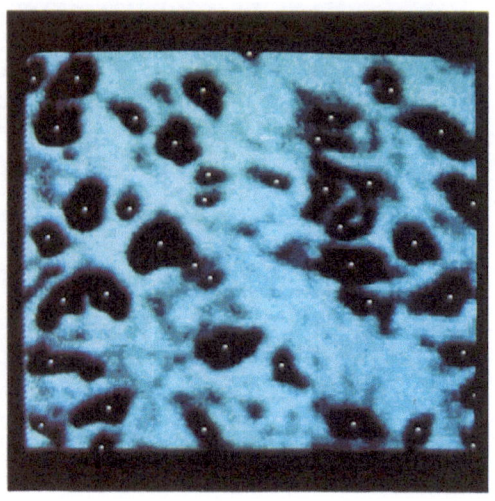

Plate 6
Cell nucleus locations
(Sternberg)

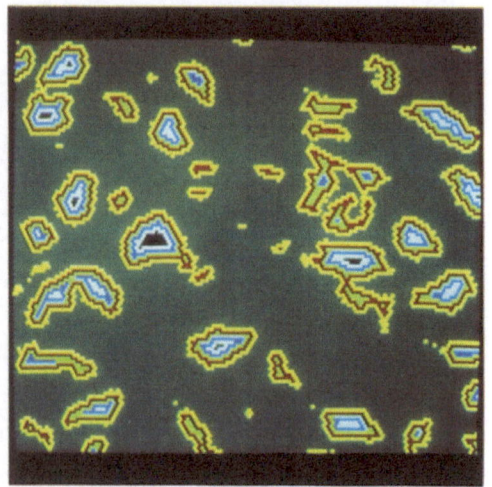

Plate 7
Nucleus topology
(Sternberg)

Plate 8
Color coding by size
(Sternberg)

Plate 9
Connectivity coding
(Sternberg)

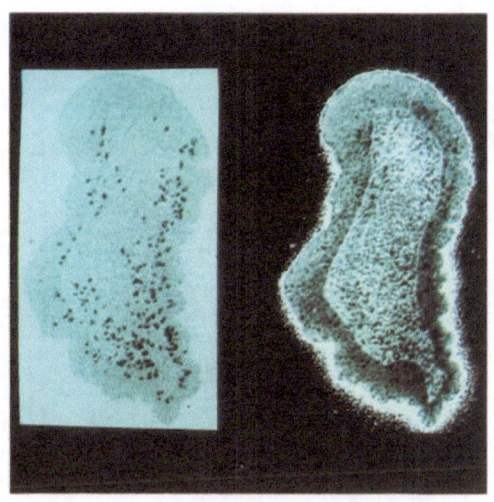

Plate 10
Autoradiography (L); Nuclei (R)
(Sternberg)

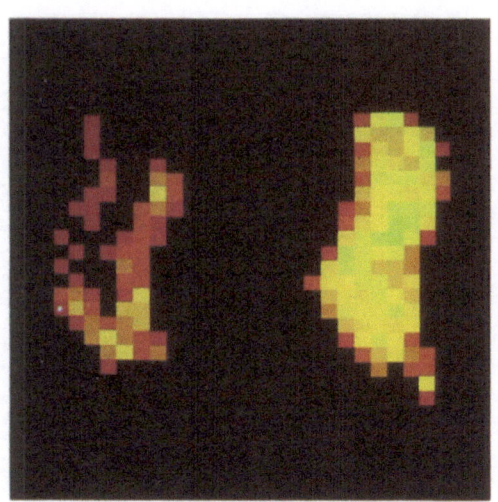

Plate 11
Recent (L) and total (R) nuclei
(Sternberg)

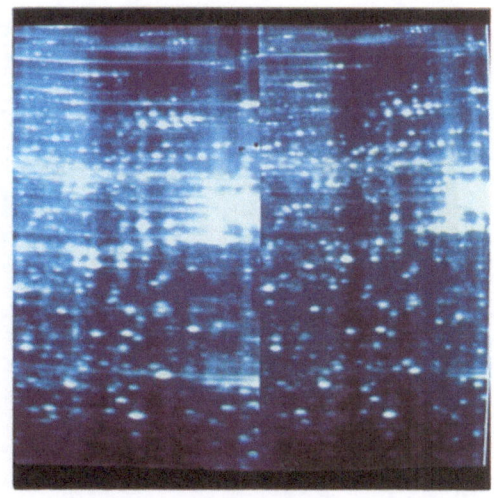

Plate 12
Two 2-D gel patterns
(Sternberg)

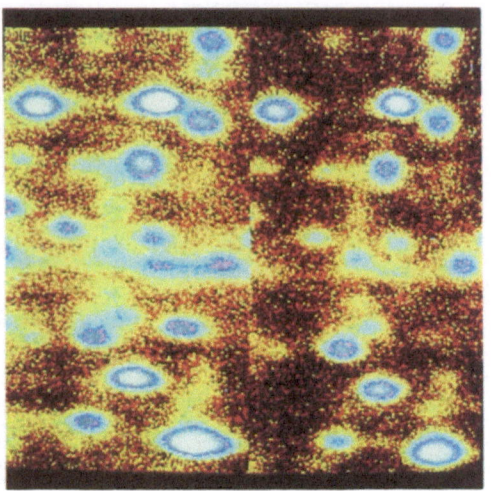

Plate 13
Detail from 2-D gel patterns
(Sternberg)

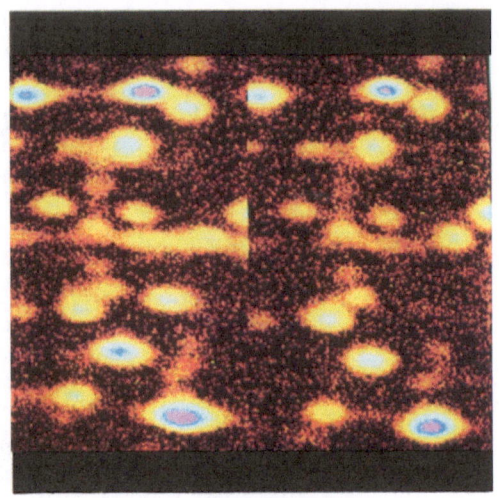

Plate 14
Background normalization
(Sternberg)

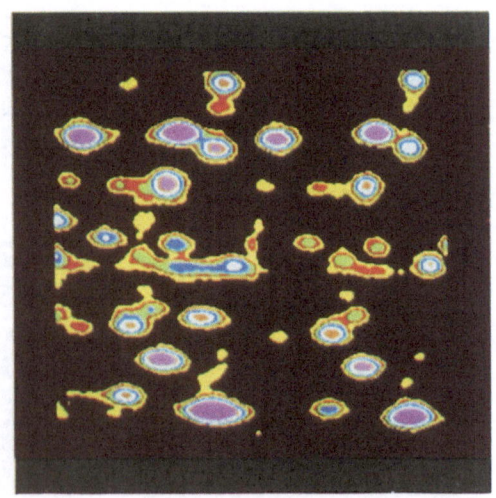

Plate 15
Boundary topology
(Sternberg)

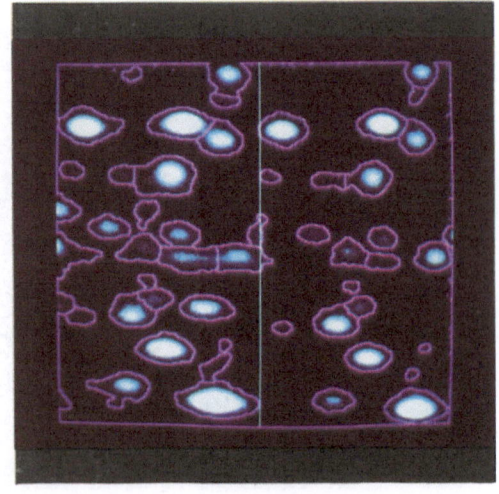

Plate 16
Protein boundaries
(Sternberg)

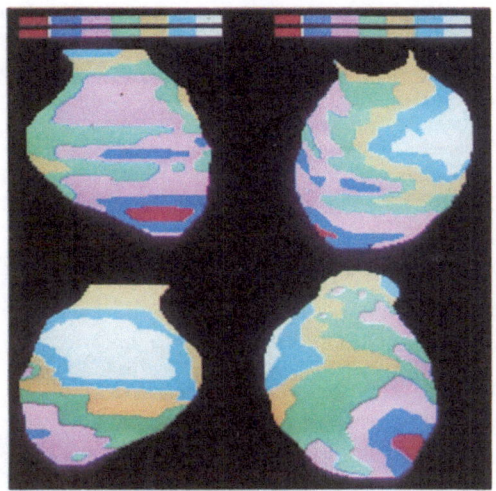

Plate 17
Color-coded left ventricle
Graphic display and composite
(Kuwahara)

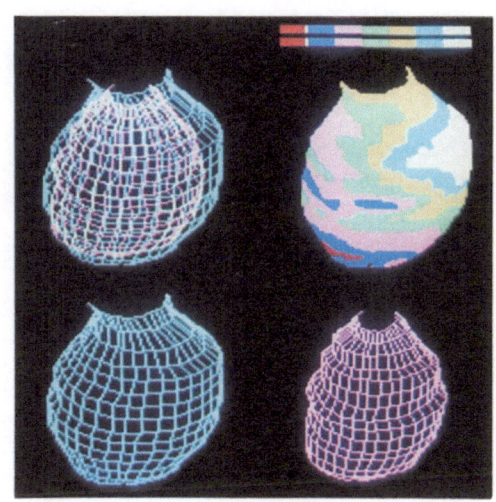

Plate 18
Color-coded left ventricle
Composite color code (L to R):
0~5,~10,~15,~20,~25,>25% shortening
(Kuwahara)

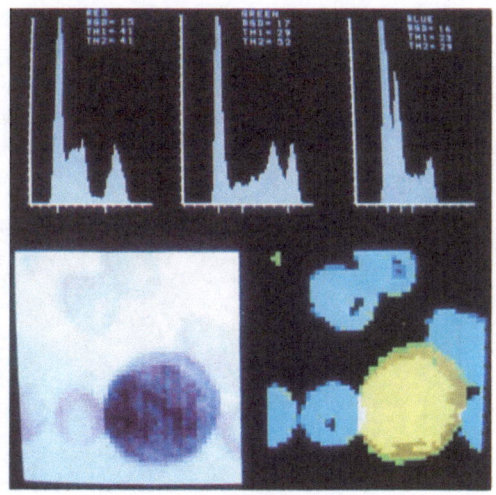

Plate 19
Color cartoon of blast
showing segmentation difficulty
(Green)

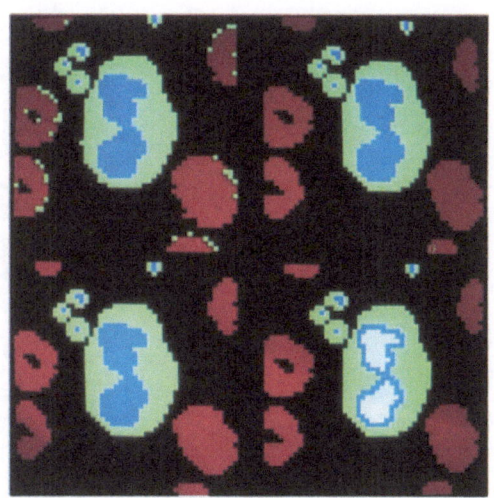

Plate 20
Three-step filtering
from raw segmentation
(Green)

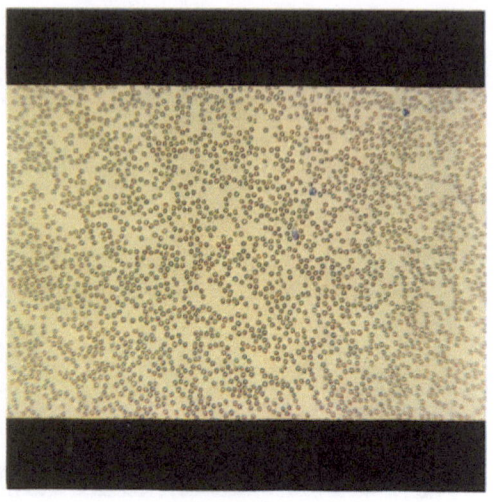

Plate 21
A diff3 blood smear 40x
(Graham and Norgren)

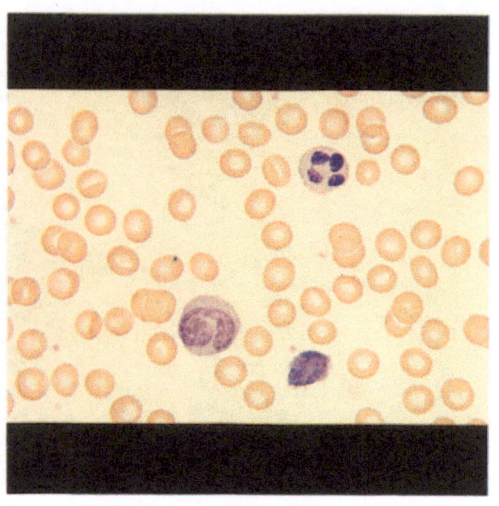

Plate 22
A diff3 blood smear 250x
(Graham and Norgren)

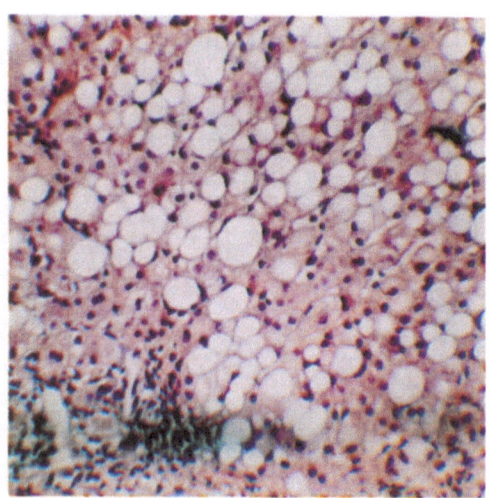

Plate 23
Fatty liver tissue
(Preston)
(Original courtesy Dicomed)

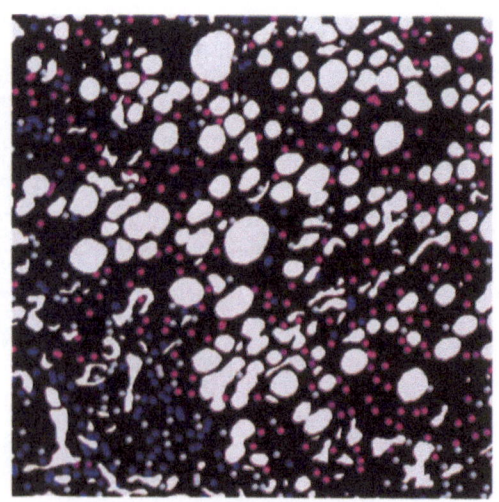

Plate 24
Cell reference map
(Preston)
(Original courtesy NASA Ames)

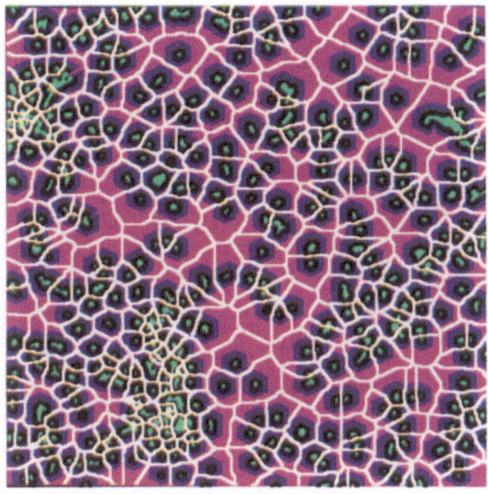

Plate 25
Growth of exoskeleton
(Preston)
(Original courtesy Dicomed)

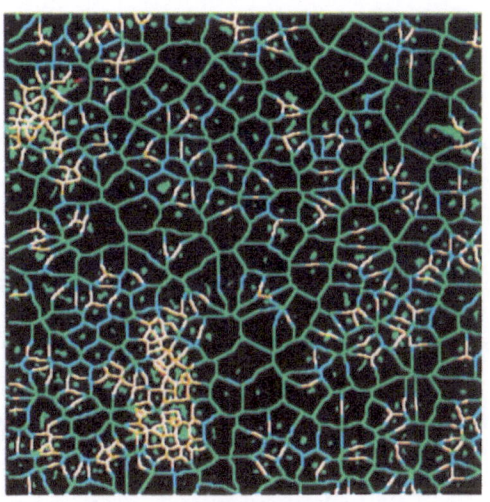

Plate 26
Exoskeleton arc formation
(Preston)
(Original courtesy Dicomed)

Plate 27
Normal kidney exoskeleton
(Preston)
(Original courtesy Dicomed)

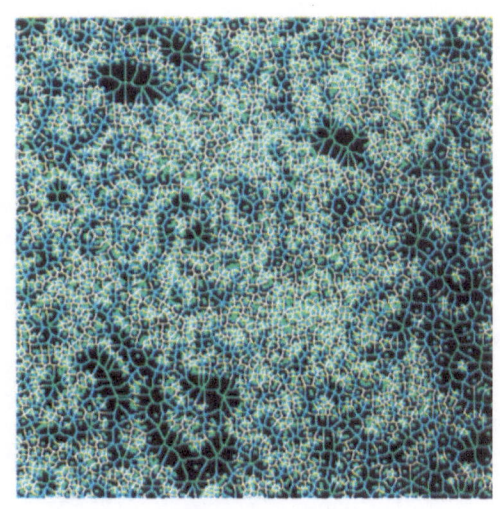

Plate 28
Abnormal kidney exoskeleton
(Preston)
(Original courtesy Dicomed)

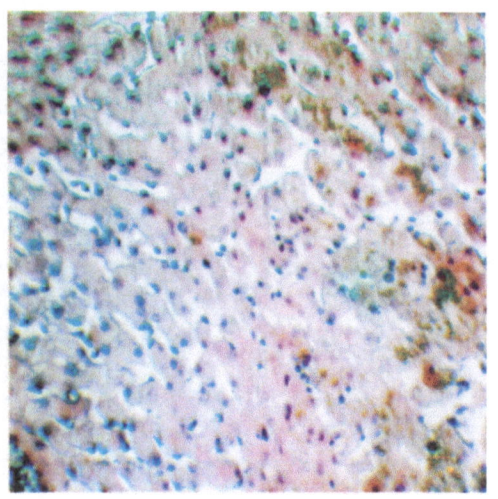

Plate 29
Normal liver tissue
(Preston)
(Original courtesy NASA Ames)

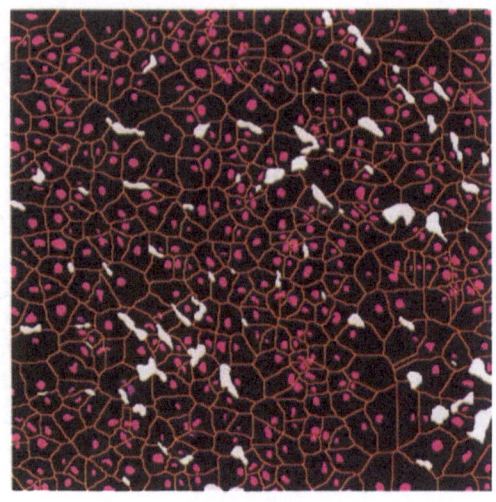

Plate 30
Normal liver tissue cartoon
(Preston)
(Original courtesy NASA Ames)

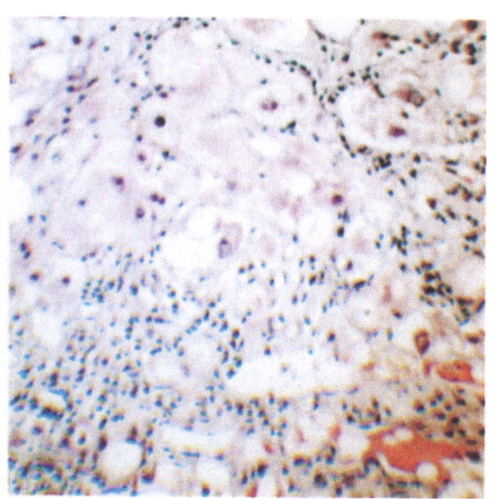

Plate 31
Abnormal liver tissue
(Preston)
(Original courtesy NASA Ames)

Plate 32
Abnormal liver tissue cartoon
(Preston)
(Original courtesy NASA Ames)

When comparing human differentials and automatic differentials, it quickly becomes apparent that humans do not perform differentials the way they claim. In theory, humans sequentially select 100 leukocytes on the blood film and classify each leukocyte according to its appearance. In fact, it appears that this is only partly true. In addition, there is apparently a concurrent evaluation of the overall appearance of the smear performed during the first part of the differential. During this evaluation process immature or abnormal cells are not reported unless they are grossly evident. During the remainder of the differential, immature or abnormal cells are only reported if the film has been judged abnormal; otherwise all cells are classified as one of the six normal types. This phenomenon became apparent when comparing human versus automatic differentials of "hospital normals." The automatic differentials revealed a significant number of atypical lymphocytes in many of these otherwise normal patients, in some cases up to 15% of the differential. However, in none of these samples were any atypical lymphocytes reported in the human differentials. Reviewing the cells classified atypical lymphs, in almost every case several humans agreed that the cells were atypical. Apparently during the hospital differential, the person performing the test judged the sample to be normal and simply classified all lymphocytes in the normal category.

6.   FUTURE TRENDS

The increase in the cell analysis rate in the ADC-500 over that previously available raises the possibility that further increases in performance might be forthcoming. Settling in the x-y stage is one major limit to system throughput. Available light is another. There are several techniques which could increase the speed of the x-y stage at the expense of increased costs. Techniques could be used to drive the stepping motors in an analog fashion so that the driven masses are smoothly accelerated and decelerated. The penalty is increased complexity and cost in the drive electronics. By using these techniques, it may be possible to more than halve the present settling time.

An increase in scan or focus system speed requires an increase in incident light or an increase in sensitivity of the sensors. An increase in light intensity by a factor of two or more using a xenon lamp would be difficult. Laser illumination is an obvious way of increasing intensity, but the color algebra method requires simultaneous illumination at specific wavelengths which may present a problem. It is possible that array sensors with greater sensitivity can be obtained.

Increasing the speed of the image analysis algorithms incorporated in the Preprocessor would be simple. A two to three times increase could be achieved by raising the clock rate and substituting higher speed logic in several circuits. One must also consider that during the move and settle cycle, cell classification is proceeding in parallel. Some refined program coding and possibly a faster computer might be necessary to match a decrease in scene cycle time. Finally, an increase in cell throughput rate could be achieved by using larger arrays to increase the resolution of the cell finder sensor and increase the scanned area of the high resolution sensors. With larger arrays, cell miscentering could be eliminated and low resolution pattern recognition could be performed to prevent the centering of many artifacts, increasing the ratio of scenes with classifiable cells. Using all these techniques, 1000 or 2000 cell differentials might be feasible with no decrease in throughput.

7.   REFERENCES

Bacus, J. W., "Erythrocyte Morphology and Centrifugal 'Spinner' Blood Film Preparations," J. Histochem. Cytochem. 22:506 (1974).

Bacus, J. W., "The Observer Error in Peripheral Blood Cell Classification," Am. J. Clin. Path. 59:223 (1973).

Bacus, J. W., Berlanger, M. G., Aggarwal, A. K., and Trobaugh, F. E., "Image Processing for Automated Erthrocyte Classification," J. Histochem. Cytochem. 24:195 (1976).

Bacus, J.W., and Gose, E. E., "Leukocyte Pattern Recognition," IEEE Trans. Sys. Man. Cyb. SMC 2:513 (1972).

Bartels, P. H., and Wied, G. L., "Future Aspects of High Resolution Analysis of Cellular Images," The Automation of Uterine Cancer Cytology (Wied, G. L., Bahr, G. F., and Bartels, P. H., eds.), Tutorials of Cytology, Chicago (1976).

Cain, R. W., and Anderson, A. C., "Parametric and Non-Parametric Analysis for Leukocyte Classification," Proc. IEEE Southeastcon 77, IEEE Catalogue No. 77 CH01233, 6:111 (1977)

Green, J. E., Analysis Method and Apparatus Utilizing Color Algebra and Image Processing Techniques. U.S. Patent 3,851,156; Nov. 26, 1974.

Green, J. E., Method and Apparatus for Dual Resolution Analysis of a Scene. U.S. Patent 3,970,841; Jul. 20, 1976A.

Green, J. E., Method and Apparatus Utilizing Color Algebra for
    Analyzing Scene Regions. U.S. Patent 3,999,047; Dec. 21,
    1976B

Green, J. E., Method and Apparatus for Dual Resolution Analysis of
    a Scene. U.S. Patent 4,061,914; Dec. 6, 1977.

Green, J. E., Method and Apparatus for Producing a Suspension of
    Biological Cells On a Substrate. U.S. Patent 4,084,902;
    Apr. 18, 1978.

Green, J. E., "A Practical Application of Computer Pattern Recog-
    nition Research. The Abbott ADC-500 Differential Classifier,"
    J. Histochem. Cytochem. $\underline{27}$:150 (1979A).

Green, J. E., "Rapid Analysis of Hematology Image Data. The ADC-
    500 Preprocessor," J. Histochem. Cytochem. $\underline{27}$:164 (1979B).

Green, J. E., "Sample Preparation Variation and Its Effects on Auto-
    mated Blood Cell Differential Analysis," Anal. & Quant. Cytol.
    $\underline{1}$:187-201 (1979)

Ingram, M., and Minter, F. M., "Semiautomatic Preparation of
    Coverglass Blood Smears Using a Centrifugal Device," Am. J.
    Clin. Pathol. $\underline{51}$:214 (1969).

Ingram, M., and Preston, K., Jr., "Automatic Analysis of Blood
    Cells," Sci. Amer. $\underline{223}$(5):72 (1970).

Kujoory, M. A., Mayall, B. H., and Mendelsohn, M. L., "Focus-
    Assist Device for a Flying-Spot Microscope," IEEE Trans. Bio-
    Med. Eng. $\underline{BME-20}$:126 (1973).

Megla, G. K., "The LARC Automatic White Blood Cell Analyzer,"
    Acta. Cytol. $\underline{17}$:3 (1973).

Miller, M. N., "Design and Clinical Results of Hematrak[R]: An
    Automated Differential Counter," IEEE Trans. Biomed. Eng.
    $\underline{BME-23}$:400 (1975).

Preston, K., Jr., "Automation of the Analysis of Cell Images,"
    Anal. & Quant. Cytol. $\underline{2}$:1-14 (1980).

Preston, K., Jr., Duff, M. J. B., Levialdi, S., Norgren, P. E.,
    and Toriwaki, J-i., "Basics of Cellular Logic with Some
    Applications in Medical Image Processing," Proc. IEEE $\underline{67}$(2):
    826-856 (1979)

Preston, K., Jr., and Norgren, P. E., Method of Preparing Blood
    Smears. U.S. Patent 3,577,267; 1971

Prewitt, J. M. S., "Parametric and Nonparametric Recognition by
    Computer: An Application to Leukocyte Image Processing,"
    Advances in Computers, New York, Academic Press (1972),
    Vol. 12, p. 285.

Prewitt, J. M. S., and Mendelsohn, M. L., "The Analysis of Cell
    Images," Ann. NY Acad. Sci. 128:1035 (1966).

Rumke, C. I., "Variability of Results in Differential Cell Counts
    on Blood Smears," Triangle 1:154 (1973).

Staunton, J. J., Clinical Spinner. U.S. Patent 3,705,048; 1972.

Young, I. T., "The Classification of White Blood Cells," IEEE
    Trans. Biomed. Eng. BME-19:291 (1972).

Young, I. T., and Paskowitz, I. L., "Localization of Cellular
    Structures," IEEE Trans. Biomed. Eng. BME-22:35 (1975).

AN APPROACH TO AUTOMATED CYTOTOXICITY TESTING BY MEANS OF DIGITAL IMAGE PROCESSING

A. Kawahara

Research Laboratory, Nippon Kogaku K.K.

Nishioi, Tokyo, JAPAN

1. INTRODUCTION

HLA antigens exist on the surface of human lymphocytes and have been widely investigated as histocompatibility antigens mainly in connection with kidney transplantation. In addition to this, it has recently been found that there are associations between certain types of HLA antigens and various diseases. The microdroplet lymphocytotoxicity test is universally accepted as the standard method for HLA antigen determination (Terasaki et al., 1978).

Although several attempts have been made to construct automatic HLA antigen testing instruments, none have been used clinically because of constant modifications of the standard testing methods as reported by Hulett et al. (1970), Bruning et al. (1972), Wilson et al. (1973). We have constructed an automatic lymphocytotoxicity testing instrument based on the present standard method. This instrument provides a specially designed image input device and an autofocusing unit for a phase contrast microscope.

In this chapter we describe the principle of the lymphocytotoxicity test and the basic idea of our instrument. We also present a comparison of machine inspection with that of human inspection.

2.   THE PRINCIPLE OF THE LYMPHOCYTOTOXICITY TEST

The procedure used in lymphocytotoxicity testing is as follows. First a suspension of lymphocytes of about 1μℓ is mixed and reacted with an equal amount of antiserum. Next rabbit complement of about 5μℓ is added to the mixture. Then eosin is used for staining and, finally, formaldehyde is added to the mixture for fixation of the reaction. The reaction is done in 60 small wells bored in the type of plastic tray shown in Figure 1.

In the case of dead cells, the lymphocytes killed by antibody in the presence of complement are stained dark. As shown in Figure 2(a), they are usually large and round. On the contrary, in case of living cells, those lymphocytes are neither killed nor stained. As shown in Figure 2(b) they are small compared with dead lymphocytes and have bright concentric rings called halos around them. Since living cells are not stained, it is necessary to use a phase contrast microscope to observe them. Clinical technologists judge the degree of positivity of each well with one glance and decide upon HLA type from the reaction pattern with respect to various antisera.

To construct an automatic inspection instrument, the following important features must be taken into account:

(2.1) Since a plastic tray has 60 wells, the automatic inspection instrument should provide for both automatic stage scanning unit and autofocusing.

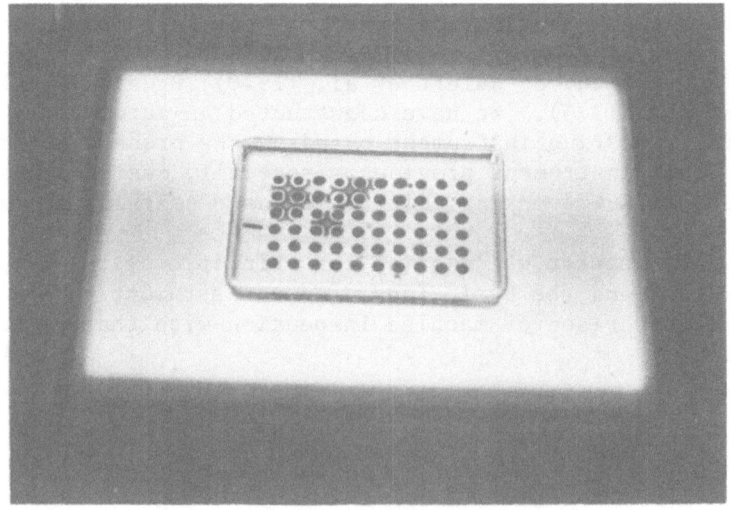

Fig. 1.   Tray used in the microdroplet lymphocytotoxicity test.

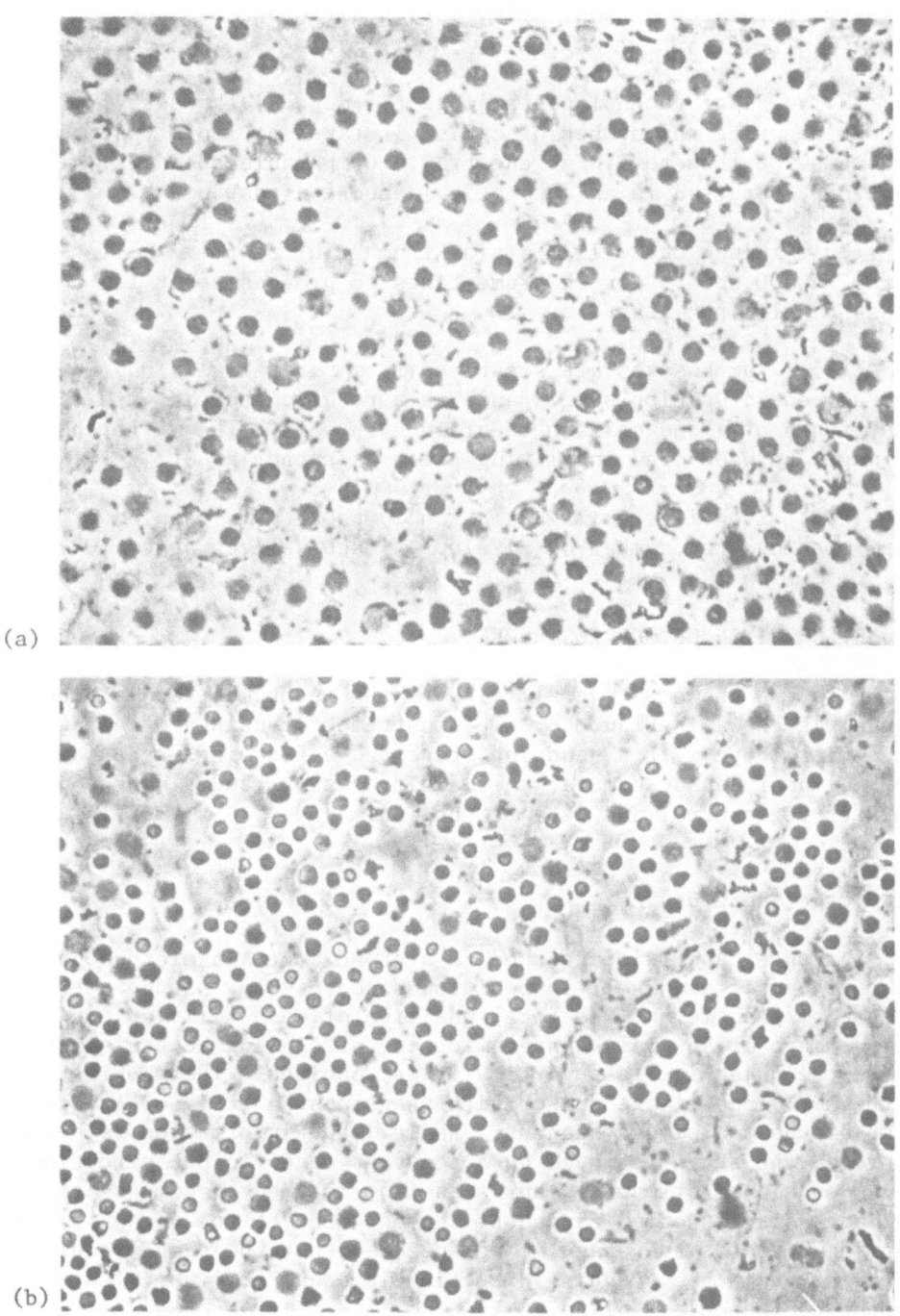

Fig. 2. Photomicrograph of cells demonstrating the lymphocytotoxicity test. (a) Dead cells and (b) living cells.

(2.2) From the practical point of view, it is sufficient to classify the degree of positivity into 5 different classes (++, +, +-, -, --).

(2.3) Since technologists judge the degree of positivity without counting the number of cells, it takes not more than 4 minutes for them to finish one tray. Comparable speed is, therefore, required in an automatic inspection instrument.

(2.4) Since one well contains from $2 \cdot 10^3$ to $3 \cdot 10^3$ cells and a tray has 60 wells, the instrument has to process about 150 thousand cells per tray.

(2.5) Lymphocytes are separated from peripheral blood in several centrifugation stages and, as a result, there are few other types of blood cells present in each well. The difference in images between dead and living cells is rather great.

(2.6) Even if the image of each lymphocyte is sampled at a 1.4μm spacing, the number of pixels for each well is about 10^6.

3.   DESCRIPTION OF OUR INSTRUMENT

Figure 3 shows an overall view of the instrument and a block diagram is shown in Figure 4. The instrument consists of an automated microscope with a scanning stage and a focusing mechanism, an MOS linear sensor camera, an image processor, and a desk-top computer for control and data processing.

3.1 Image Input Device

A stage scanner is attached to a phase contrast microscope in order to center each well in position successively. Since a tray is made of plastic and the focal position of each well varies over a range of several hundred microns, a focusing mechanism which drives the objective lens vertically is required. The combination of the MOS linear image sensor with the scanning stage constitutes a two-dimensional image scanner. While image information is being recorded, the stage is moved at a constant speed in one direction. The video signal of every other scan line is used to compensate for image deterioration caused by the movement of the sensor. The scanning speed of the stage is selected so that the sampling pitch of the stage in the scanning direction is equal to 1.4μm with respect to the object plane.

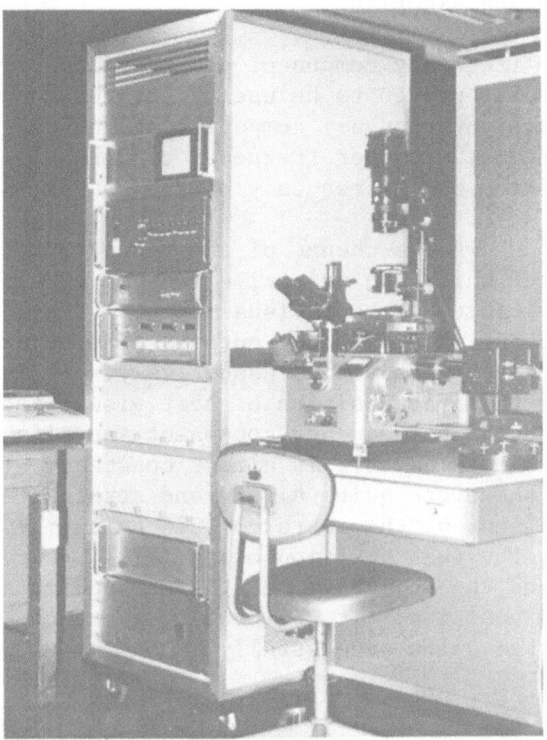

Fig. 3. Overall view of the instrument for automatic reading of trays used in the lymphocytotoxicity test.

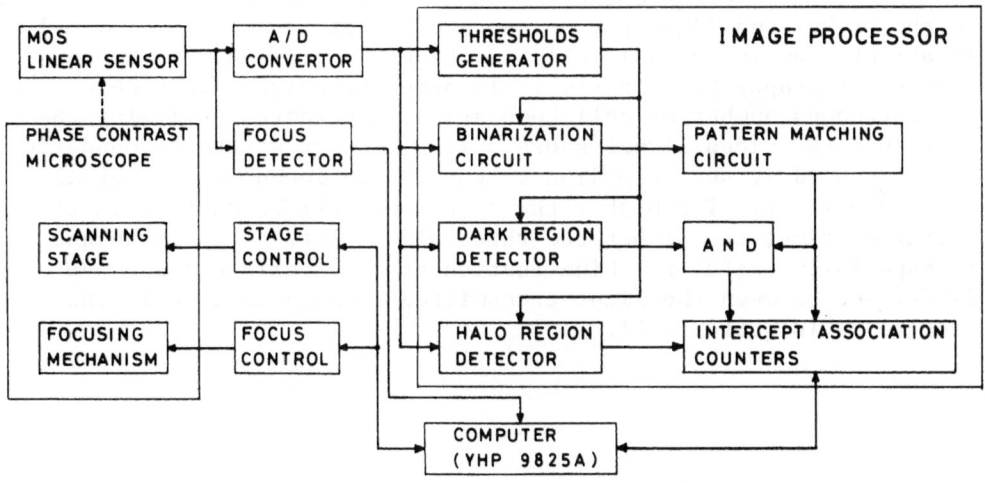

Fig. 4. Block diagram of the instrument.

3.2 Autofocusing Algorithm

The high frequency component of the video signal from the linear sensor has proven to be useful for autofocusing. In order to obtain the high frequency component an analog bandpass filter was constructed, the center frequency of which is somewhat lower than one half the clock frequency of the sensor.

Figure 5 shows the change of the rms output from the analog bandpass filter while the objective lens moves along the optical axis in the vicinity of best visual focus. Figures 5(a) and 5(b) correspond to the change of the outputs from several positive wells (++) and negative wells (--), respectively. It is apparent from these figures that the position of best visual focus and the peak position of the high frequency component are slightly different and the offset between them is almost constant. The autofocusing algorithm is therefore quite simple and straightforward. While the objective lens of the microscope is being moved along the optical axis, the peak position of the high frequency component is detected and the objective lens is stopped at a constant offset from the peak position. It was confirmed experimentally that this algorithm worked quite successfully.

3.3 Thresholding

The conventional constant threshold technique for binarization is not applicable in our case because the background light is not uniform over the whole field of view. We therefore employed a dynamic threshold in which the averaged local video signal is used to determine the instantaneous threshold for binarization. The averaging time should be short enough to compensate for fluctuations in the background light but not so short that the threshold level is affected by individual cells in which case it would be appropriate for proper binarization. In order to solve this problem the instantaneous number of cell boundaries is detected utilizing the fact that the video signal shows a large change at these boundaries. The threshold is set according to the instantaneous number of all cell boundaries. The higher the number of cell boundaries in the averaging time, the higher the threshold is set. This technique is especially useful for binarization of dead cells because the difference between the light transmitted through dead cells and background is quite small.

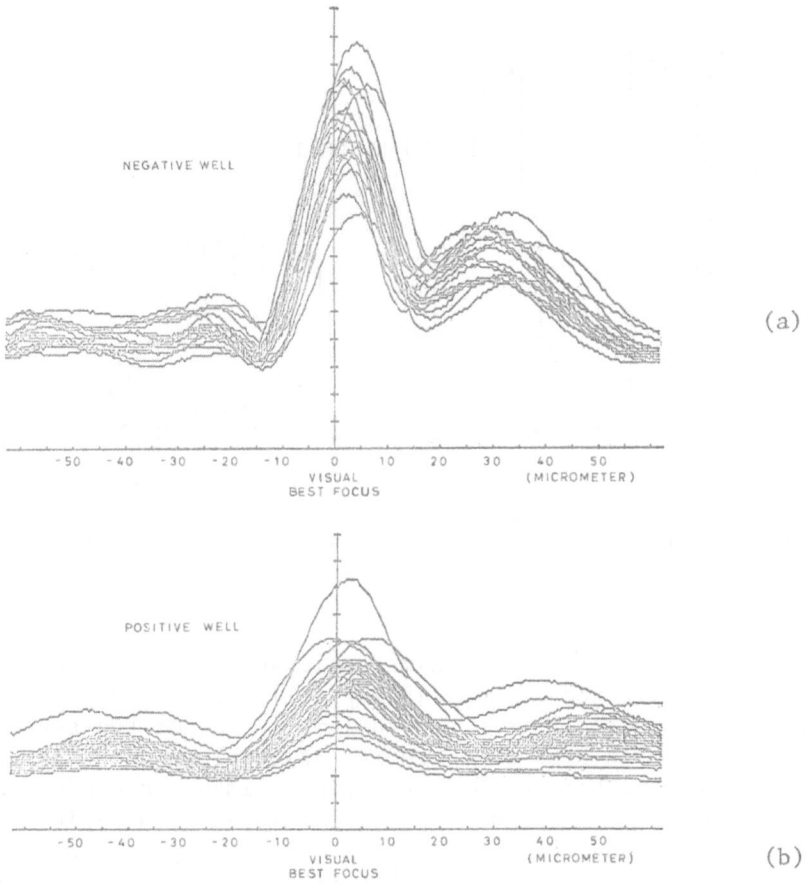

(a)

(b)

Fig. 5. The rms outputs of the analog bandpass filter for
(a) a positive well and (b) a negative well.

3.4 Binary Pattern Matching

The most important discrimination parameter between dead and
living cells is cell size. The sizing of cells is performed on
binary images generated by thresholding as described in Section 3.3.
Figure 6 shows the area distribution of dead and living cells for
the lymphocytotoxicity test, where cells are dispersed and isolated from each other. Clearly dead cells are usually larger than
living ones.

In order to classify dead and living cells on the basis of
cell size, we face two problems: (1) some living cells are larger
than dead ones and (2) many cells lie so close together that several
cells may appear to connect. Since we classify the degree of

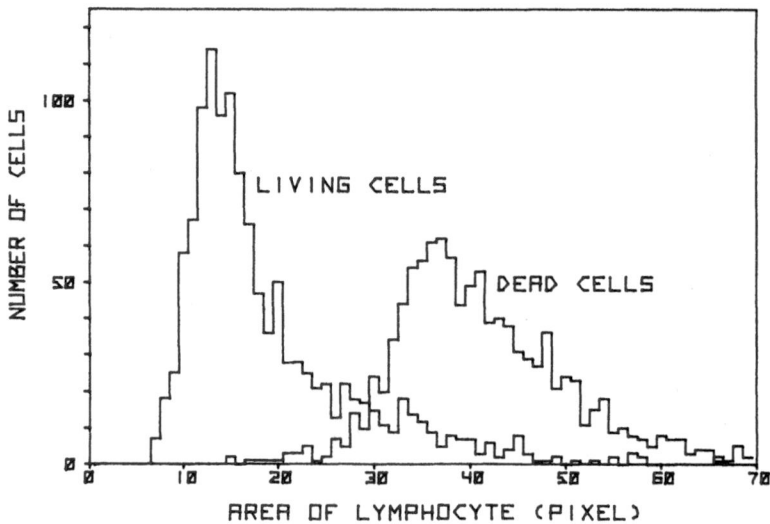

Fig. 6. Area distribution of dead and living lymphocytes in the cytotoxicity test.

positivity into five approximate classes, performance is not seriously affected if some living cells are judged as dead cells and vice versa. Therefore, problem (1) is not too serious. An algorithm which overcomes problem (2) employs the well known "shrinking" method in which the binary image is eroded in order to cut connectivity. To do this, local image information must be first stored and then the image is eroded by means of repetitive, parallel local operations. This method takes too much time and is inadequate for real-time image processing. As an alternative we have devised a binary pattern matching technique especially useful in overcoming the connectivity problem in real-time processing. This technique is described below.

Two different types of templates (Figure 7) are used in our instrument for binary pattern matching. One is used to detect large cells (i.e., positive cells) and the other to detect all cells. The size of these templates is determined by the area distribution of actual lymphocytes. The procedure for removing apparent connectivity using these templates is as follows. The original binary image is scanned by all the templates, the center pixel of the template window becomes 1 in the output image only when all pixels of any template are 1. Once the entire original binary image is scanned by the templates, cells are well isolated in the newly generated output image as compared with the original binary image. An example of the newly generated binary image of a positive well is shown in Figure 8. We see from the figure that all cells are well isolated.

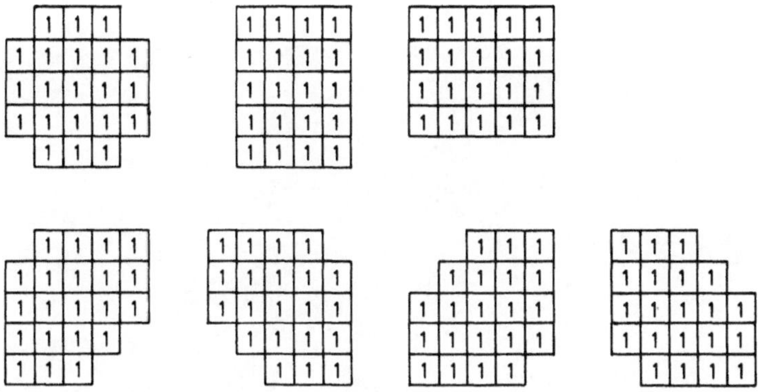

(a) Seven different templates used to detect large cells.

A_1	A_2	A_3	A_4
A_{12}	1	1	A_5
A_{11}	1	1	A_6
A_{10}	A_9	A_8	A_7

(b) Template used to detect all cells.

Fig. 7. Illustration of the template patterns used in binary pattern matching. Templates in (b) involve all the patterns, in which inner four elements have value 1, and the outer twelve elements satisfy the relation

$$\sum_{i=1}^{12} A_i \geqq 6 \quad (A_i = 0 \text{ or } 1)$$

3.5 Cell Counting

It should be noted that newly generated binary image produced by pattern matching is cleaner than the original binary image. Small noisy spots in the image are removed and all the cells are well isolated. The intercept association counting method is suitable for counting the number of convex patterns in a binary image as described by Taylor (1954) and Preston (1976). We therefore have developed our special purpose processor based on this method and the pattern matching method mentioned in 3.4.

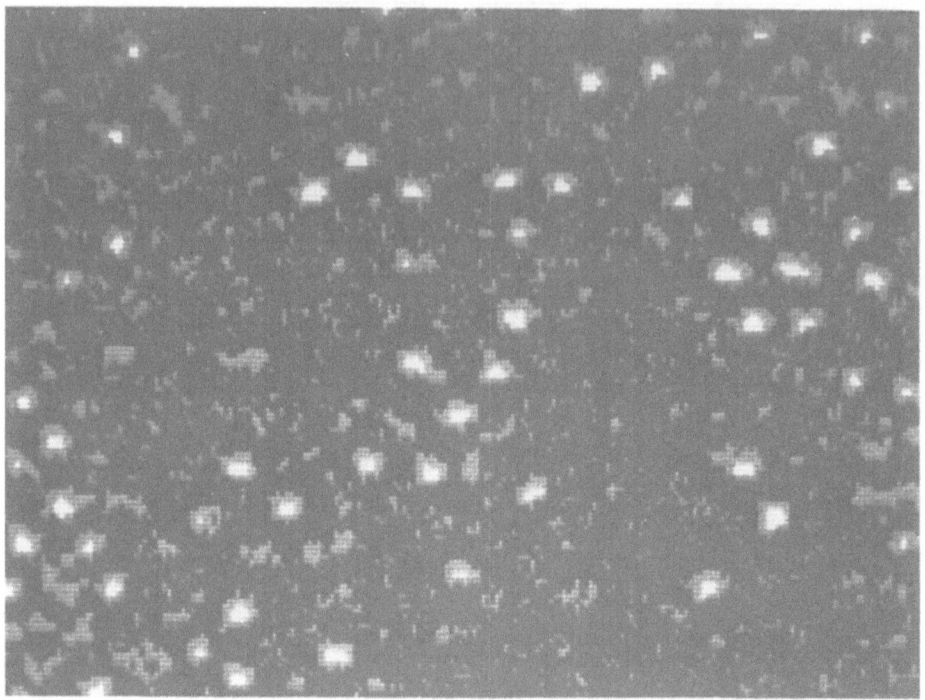

Fig. 8. Binary image patterns processed by the binary pattern
matching technique to detect the large cells of a positive well.

4.   IMPORTANT PARAMETERS TO DETERMINE POSITIVITY

    Cell size, cell density, and halo are the parameters which we
use in determining the positivity of the contents of each well.

4.1 Cell Size

    Cell size is the most important parameter for judging the
positivity of a well. All the cells detected in a well are classified according to their size. The number in each class is counted.
The area distribution can be estimated by the number of cells in
each class. Since counting is done using the newly generated
image each cell is smaller than in the original image. By counting
inside the original pattern, this enables us to combine size
parameters with the other parameters mentioned below in order to
judge comprehensively whether or not a cell is living.

4.2 Density

The density parameter is also important. Since the central
portion of a living cell is often darker than that of a dead one,
the threshold level of binarization used to separate living cells
from dead cells by their density should be near the average brightness of dark living cells and light dead cells. The binary signal
corresponding to the density of the central portion of a cell is
delayed and superposed on the counting pulse, which is generated
below to the right hand side of an original cell. A simple AND
operation between these two signals extracts only dark cells.

4.3 Halo

Several concentric rings appear on the image when we observe
living cells through a phase contrast microscope. These halos
appear only around living cells and can be used as a positivity
measure. In order to detect halos, an original image is first
differentiated and converted into a binary image by thresholding.
By choosing the proper threshold level only halo regions are
extracted and counted.

5. EXPERIMENTAL RESULTS

Among actual specimens we selected only good ones containing
no air bubbles in their wells and inspected them automatically.
Figures 9(a) and 9(b) show the relationships between various
parameters mentioned in Section 4. These charts are only examples
from tremendous amounts of data. In the figures, $N\ell$, Ns, Nb and
Nh stand for the number of large cells, total cells, dark cells,
and halos, respectively. We see from the figures that cell size
and transmittance are highly correlated and that small cells are
almost always dark cells. Also halo and cell size are highly
correlated and cells having halos are usually small cells. The
reason why Nb and $N\ell$ are divided by Na is that, by doing so, the
relationship between each of these two parameters becomes insensi-
tive to the total number of cells. By dividing Nb by $N\ell$ instead
of Ns, the relationship between the number of cells having halos
and the number of the large cells becomes more clear.

The positivity of a well is judged as follows. First the
above mentioned quantities, such as $N\ell$, are counted by our instrument and then ratios, such as $(N\ell/Ns)$, are calculated. By con-
sulting either of two charts, the positivity of a well is finally
judged. The reliability of this judgement using either chart is
almost the same and, according to Bayes' rule, the error probability of 0%–not 0% is about 5% and that of 100%–not 100% is about
2.5%.

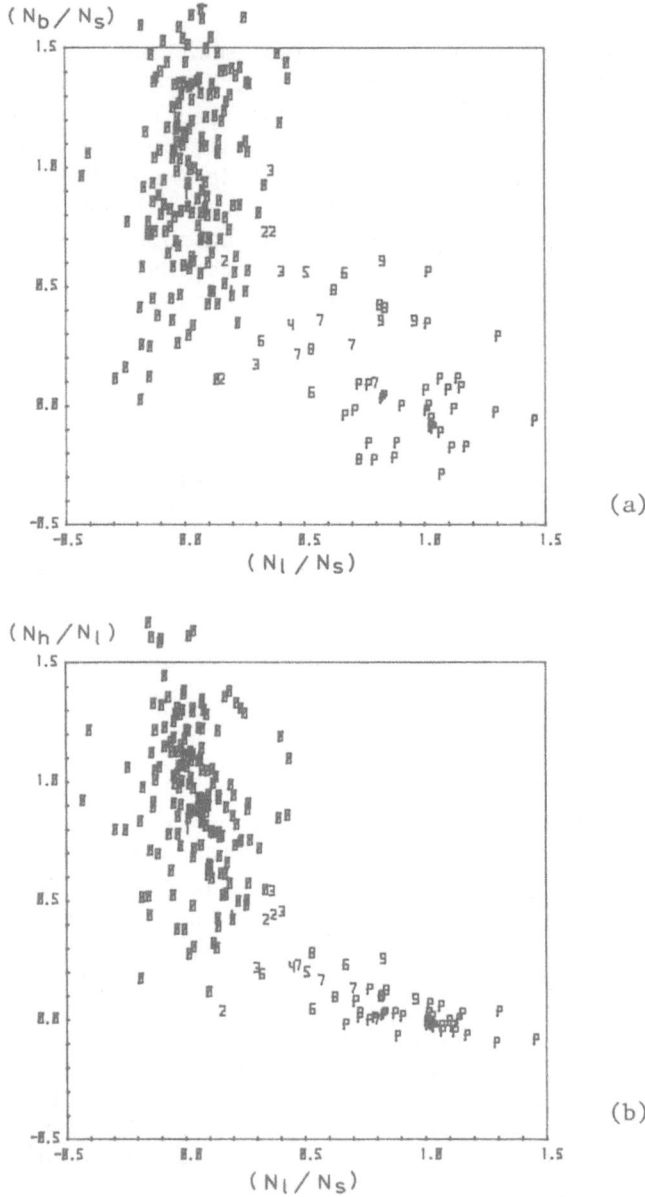

Fig. 9. Relationship between various effective parameters used
in judging the positivity of a well in the cytotoxicity test:
(a) density versus cell size and (b) halo versus cell size. In
these figures Nℓ, Ns, Nb, and Nh stand for the number of large
cells, total cells, dark cells, and halos respectively. The
plotted digits 0–9 and P are the degree of positivity read by
technologists (0%–90% and 100%).

A comparison of the results obtained by the instrument compared with that obtained by technologists is shown in Figure 10. This figure shows that the technologists' inspection may be replaced by machine inspection.

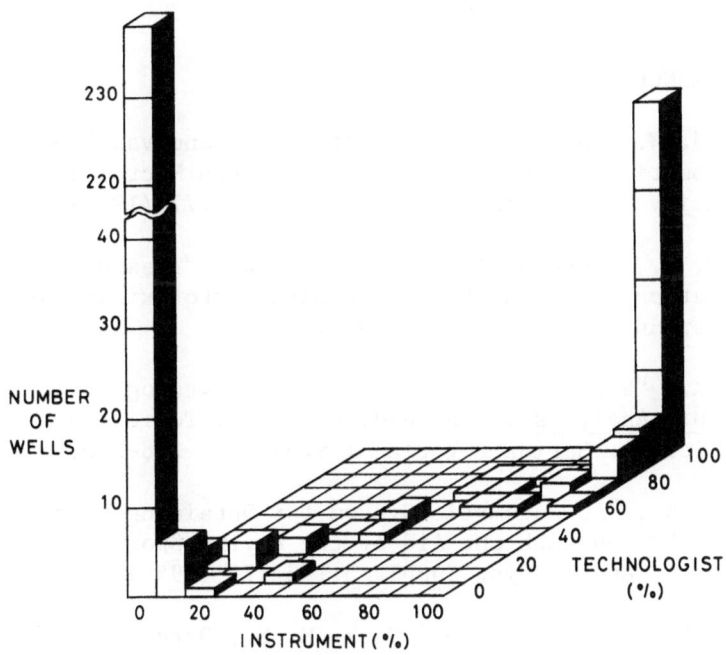

Fig. 10. Comparison of the degree of positivity as computed by the instrument and as read by technologists.

6. SUMMARY

For the purpose of automatic inspection of the microdroplet lymphocytotoxicity test, we have developed an image scanner and a real-time processor. It was found that the combination of a linear MOS image sensor with a mechanical stage scanner is useful as a microscopic image scanner especially when large areas must be scanned continuously with high resolution. The intercept association counting technique of Taylor and Preston was used effectively to count cells after removal of apparent cell-to-cell connectivity by means of our method of binary pattern matching. It was also found that these techniques were successfully applicable to the analysis of images produced from lymphocytotoxicity tests and in the future we believe that the technologists' inspection may be replaced by machine inspection.

7.   ACKNOWLEDGEMENTS

The author wishes to express his thanks to N. Fujii and
M. Sawada of Nippon Kogaku K.K. for their technical assistance.
Thanks are also given to Prof. K. Tsuji and Assoc. Prof. H. Inoue
of Tokai University for their suggestions and for the preparation
specimens.

8.   REFERENCES

Bruning, J. W., Douglas, R., Scholtus, M., and van Rood, J. J.,
    "Automatic Reading and Recording of the Microlymphocyto-
    toxicity Test," Tissue Antigens 2:473-477 (1972).

Hulett, H. R., Coukell, A., and Bodmer, W., "Tissue-Typing Instru-
    mentation Using the Fluorochromatic Cytotoxicity Assay,"
    Transplantation 10(1):135-137 (1970).

Preston, K., "Digital Picture Analysis in Cytology," in Digital
    Picture Analysis (Rosenfeld, A., ed.), Topics in Applied
    Physics 11, Springer-Verlag (1976), pp. 248-250.

Taylor, W. K., "An Automatic System for Obtaining Particle Size
    Distribution with the Aid of the Flying Spot Microscope,"
    Brit. J. Appl. Phys. Suppl. 3:173-175 (1954).

Terasaki, P. I., Bernoco, D., Park, M. S., Ozturk, G., and
    Iwaki, Y., "Microdroplet Testing for HLA-A, -B, -C, and -D
    Antigens," Am. J. Clin. Path. 69:103-120 (1978).

Wilson, M., Dardano, J., and Rothschild, H., "An Automated, Rapid,
    and Quantitative Microassay for Histocompatibility Testing,"
    Transplantation 16(5):408-414 (1973).

THE diff3[T.M.] ANALYZER: A PARALLEL/SERIAL GOLAY IMAGE PROCESSOR

M. D. Graham[+] and P. E. Norgren[#]

[+]Coulter Electronics, Inc., Concord, Massachusetts USA

[#]The Perkin-Elmer Corp., Norwalk, Connecticut USA

1. INTRODUCTION

As part of the diff3 System, the diff3 analyzer represents a practical application of real-time image processing to an important clinical need, automation of blood-smear analysis. To accomplish this task, the analyzer is supported by a complete complement of sample preparation reagents and instrumentation, including an Ames Company stainer and a proprietary diluter/spinner; design details for the latter are available [Maher and Pirc (1977a, 1977b, 1978)]. The sample preparation methodology has been evaluated by Nourbakhsh et al. (1978); it produces Wright's stained blood smears, on standard microscope slides, which are of consistently high uniformity and quality (Color Plates 21 and 22).*

The diff3 analyzer accepts a cassette holding 14 such slides and, if none require reviewing, can automatically process the full load without interruption. If a given specimen is normal, leukocytes will be classified (into segmented or band neutrophils, basophils, eosinophils, lymphocytes, and monocytes); the erythrocytes will be evaluated for normalcy of size, shape, and color; and the platelet count will be estimated. The results are printed out for each slide, and the analyzer signals the operator when it has completed the cassette. Should the specimen be abnormal, nucleated erythrocytes, immature granulocytes, blasts or atypical lymphocytes may be found; for these, as well as for occurrence of normal cell types at greater than normal rates, the analyzer indicates that the smear should be reviewed. It can also be set to flag specimens with abnormal erythrocyte morphology or platelet estimates.

---

diff3[T.M.] is a registered trademark of Coulter Electronics, Inc.
*The color plates will be found following page 144.

Because criteria for normalcy in blood-smear data will vary
from institution to institution, flexibility in flagging thresholds
is provided for all cell classes except nucleated erythrocytes and
blasts, whose thresholds are set at one percent. But the young cell
forms may occur at low rates in otherwise normal specimens, and to
prevent excessive review the operator can select either of two Context Filter settings. If the leukocyte differential is otherwise
normal, one cell classed as a blast will be ignored in the first,
while in the other the analyzer ignores one cell classed as a blast,
an immature granulocyte, or unclassifiable. In both these settings
the analyzer will ignore a single erythrocyte identified as nucleated, if the erythrocyte morphology is normal. Similarly, the plane
separating segmented from band neutrophils, and atypical from normal
lymphocytes, is placed under operator control, so that continuity
with existing data bases need not be affected.

Regardless of the criteria used to establish need to review a
given specimen, two modes are provided for its implementation. In
the Full Review Mode, results for each blood smear are displayed on
an internal monitor prior to printing, and the operator can select
any classification category for review. Cells classed within that
category will be relocated, reclassified, and presented for visual
examination; the operator can elect to change the classification of
any given cell before the final report is printed. Alternatively,
in the Conditional Review Mode the analyzer proceeds through the
cassette doing a complete analysis and printout for each smear, until it encounters one requiring review. It then switches to the
Full Review Mode and signals the operator, who reviews the appropriate category before initiating printout and re-entry into the Conditional Review Mode. In all operational modes, however, the result
is a printed report suitable for inclusion in the patient's file.

Performance of the diff3 System in clinical settings has been
evaluated, by Gilmer (1978) and Schoentag and Pedersen (1979) for
prototypes and by Nosanchuk et al. (1980) for an early production
system. Preston et al. (1979, 1980) and Norgren et al. (1980) have
discussed theoretical aspects of image processing in the analyzer;
the first authors give comparative data on all commercial automated
differential analyzers manufactured in the United States, and the
latter treat systems considerations for the diff3 more completely
than will be possible in this summary.

Here, the internal structure through which the diff3 analyzer
achieves the functional characteristics outlined above will be presented, with the intention of illustrating the generality and flexibility gained by application of real-time, parallel image processing
techniques. Examples drawn from blood-smear analysis will be used to
demonstrate analyzer characteristics applicable to a broad range of
biomedical problems requiring analysis of slide-mounted specimens.

2.   FUNCTIONAL ORGANIZATION

The need for reproducible sample preparation cannot be over-emphasized, but even with excellent control over specimen quality, totally automated analysis of slide-mounted material remains a complex task (Table 1). Accordingly several approaches have been followed in commercial, automated blood-smear analyzers [Green (1979), Megla (1973), Miller (1976)]. That taken in designing the diff3 analyzer has emphasized automation of the routine, repetitive operations, while giving the operator control in the crucial, often judgemental decisions whose unsupervised automation may be neither desirable nor cost-effective. Deliberate effort has been spent to avoid limiting the resulting hardware to blood-smear analysis.

Table 1 - Operations necessary for automated analysis of slide-mounted specimens. Implementation in the diff3 analyzer is indicated through references to Figure 1; control functions of the minicomputer are omitted. Comments related to analysis of blood smears are enclosed in parentheses.

Operation	Implementation in diff3
1. Load and oil slide.	Automated microscope.
2. Acquire approximate focus.	Automated microscope.
3. Find object (cell nucleus).	Automated microscope and finder logic in Golay image processor.
4. Acquire optimal focus.	Automated microscope and histogrammer in image processor.
5. Center object in view (WBC).	Auto. microscope; histogrammer and GLOPR of image processor.
6. Scan image field.	Automated microscope.
7. Segment image field (background, RBC, platelet, WBC nucleus, WBC cytoplasm).	Histogrammer and GLOPR of Golay image processor.
8. Extract features (color, size, shape, texture).	Histogrammer and GLOPR of Golay image processor.
9. Classify object (cell).	Minicomputer.
10. Repeat 3-9 for n objects.	Auto. microscope, Golay image processor, and minicomputer.
11. Format and display data.	Minicomputer and data monitor.
12. Output (reviewed) results.	Minicomputer and printers.
13. Off-load slide.	Automated microscope.
14. Index to next slide.	Automated microscope.
15. Repeat 1-14 for next slide.	Complete analyzer.
16. Signal end of cassette.	Automated microscope.

The internal structure chosen to implement these design goals
is shown in Fig. 1. As will be discussed in Sections 3 and 4, both
the automated microscope and the Golay image processor are specially
designed devices of general applicability, and both act as special-
purpose peripherals to the Data General Nova minicomputer, itself a
general-purpose 16-bit computer. Other peripherals are more familar:
Two joysticks, for manual control of the automated microscope's
focus and stage drives; a keyboard and a data monitor, for operator
interaction; a second black-and-white monitor, for display of image
scans and their processing; two printers for report output; an audio
alarm, for signaling; and a tape drive, for loading operational and
diagnostic software and for temporary storage of low-speed data.
Provision is also made for control of the analyzer by, or data pas-
sage to, an external computer via a standard RS-232 port. As a result of this internal organization, operational characteristics of
the diff3 analyzer are largely determined by the programs loaded
into the minicomputer's memory via the tape drive; as will be seen,
choice of the objective and color filters used in the microscope
are the major hardware concessions to the blood-smear application.

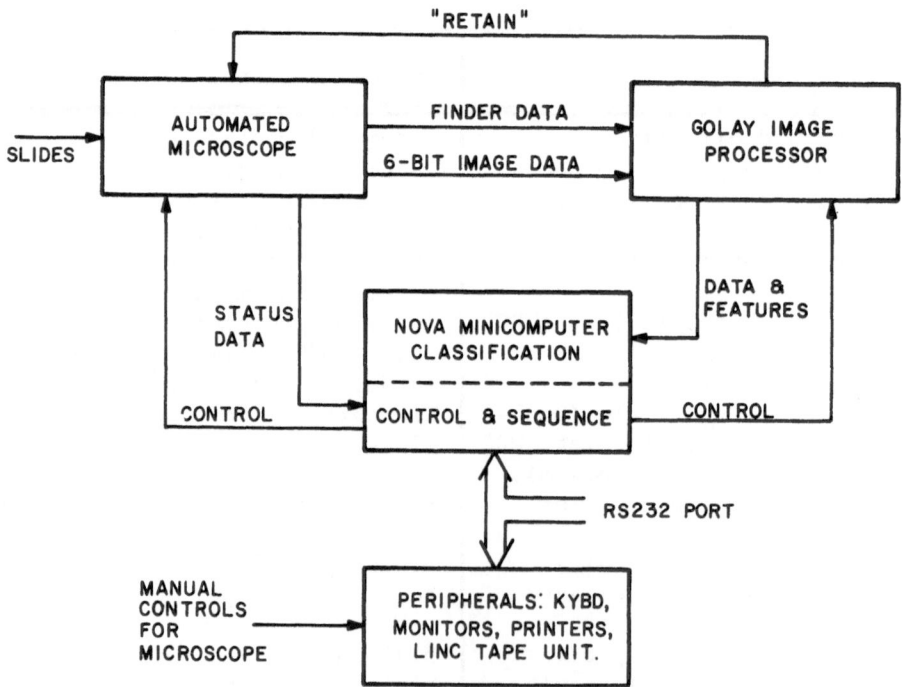

Fig. 1. Internal structure of the diff3 analyzer. Although not
indicated, the Golay image processor includes logic related to object finding and image histogramming, in addition to that for parallel image processing (GLOPR).

Program tapes presently available for diff3 analyzers include separate algorithms for analysis of Wright's stained blood smears, for counting reticulocytes in New Methylene Blue stained smears and for field-diagnostic use; experimental tapes have been developed to expand analyzer capabilities into other areas of developmental or research interest. Although detailed implementation of individual operations reflect specific applications, the sequence of Table 1 applies to most, and analysis of blood smears is illustrative.

After a cassette is placed on the autoloader of the automated microscope, the operator uses the keyboard both to interactively select an analysis protocol from a menu displayed on the data monitor and to initiate the analysis sequence. The analyzer then automatically completes Steps 1 through 10 of Table 1, if operated in the Full Review Mode, or Steps 1 through 16 for slides not needing review, if operated in the Conditional Review Mode. If in the latter case a slide is encountered which requires review, the analyzer switches to the Full Review Mode at Step 11, stops and signals the operator, who reviews the appropriate category and returns the analyzer to its Conditional Review Mode. This approach automates all but the unsupervised printout of data for the crucial slides which fail standards for normalcy and provides two methods for operator supervision of such data, either slide-by-slide as each slide is completed or as each such slide occurs. It is believed that this is an effective and a cost-effective approach to blood-smear analysis, using as it does the experience of the operator to great advantage. Similar advantages can accrue in other analyzer applications.

Interactions of the several internal subsystems of Figure 1 are broadly indicated in Table 1; more-detailed treatment will be developed along with descriptions of the automated microscope and the Golay image processor in Sections 3 and 4, respectively. First, however, its central importance in the classification operation, as well as in overall analyzer control, requires a brief description of the minicomputer.

The Data General Nova computers are general-purpose 16-bit devices containing four accumulators and operating from stored programs. Their basic instruction set permits fixed-point arithmetic and logic operations between accumulators, transfer of operands between accumulators and main memory, transfer of program control, and input/output operations. These last are supported by a bus structure consisting of the 16 data lines in addition to six lines for peripheral selection; interrupt capability is also provided. In standard configuration, main memory is limited to 64 Kbytes and can be directly or indirectly addressed; a peripheral may also access memory directly via a high-speed data channel. These computer characteristics allow real-time classification of cells using the image data and features derived by the Golay image processor.

3.   AUTOMATED MICROSCOPE

The automated microscope implements all mechanical and sensing
functions necessary for total automation of Steps 1 through 6 of
Table 1. After a loaded cassette is placed in the autoloader, the
three-axis mechanical stage is offset in the x- and y- directions
so that it can accept a slide, the slide is loaded (and oiled), and
the stage returned to its home position, all under control of the
minicomputer. In the home position, stored data related to stage
characteristics are used to move the stage in the z-direction, so
that coarse focus is obtained and the slide is within range of the
fine-focussing system. Cells are then located by driving the stage
while finder logic in the Golay image processor examines the output
of a line scanner in the automated microscope. When a suitable object is found, the stage is stopped, and by use of a rotating mirror, the object is centered under control of the image processor's
GLOPR. Optimal focus is acquired by piezo-electrically moving the
slide in the z-direction until contrast is maximized in the output
of the television camera used for data acquisition; the histogrammer in the image processor controls final slide position. High-
resolution images in red and green are scanned and digitized, for
analysis by the Golay image processor. The microscope continues
its operational cycle, until the specified number of objects has
been found and the slide off-loaded, while the image processor and
minicomputer complete the intermediate steps of image segmentation,
feature extraction, classification, and data formating. At any
point during the analyzer's cycle the operator can view the area
of the slide being scanned, using the same objective through which
data is being acquired. Operational limits of the automated microscope are defined by characteristics of its hardware, which will
now be examined in detail.

3.1 Optics

Optics used in the diff3 analyzer are premium commercial components intended for use in laboratory microscopes; the only items
related to the blood cell application are the color filters, which
have been selected for use with Wright's staining, and to a lesser
extent the objective. The optical layout (Figure 2) has the slide
illuminated from below by a 60-watt tungsten lamp and a dry 0.9NA
condenser. The objective is a 40X, 1.0NA oil-immersion planapoch-
romat suitable for use with both coverslipped and uncovered slides.
The condenser and objective are fix-mounted, properly focused and
set for Kohler illumination. Correct object focus is established
by moving the slide along the z-axis, so that the slide surface is
kept coincident with the common focal plane of the condenser and
objective; as was noted above, focussing is accomplished through
use of independent coarse- and fine- motion systems.

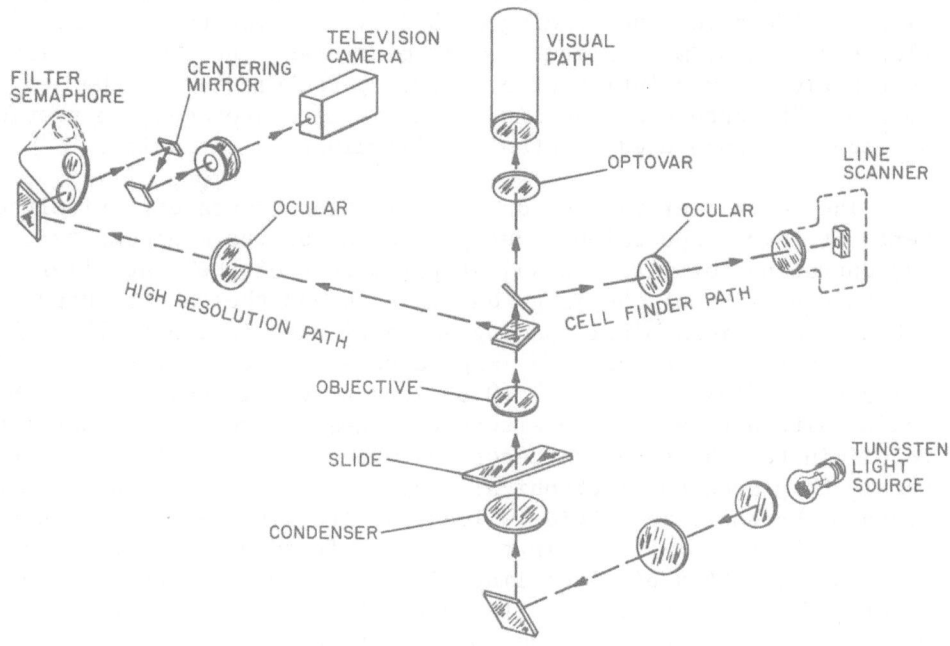

Fig. 2. Optical paths in the automated microscope.

As shown above, the image bundle formed by the objective is
divided into three paths, the first of which allows the operator
to examine the objective field-of-view at any time. The visual path
includes a standard binocular body, with 12.5X widefield oculars,
mounted above a variable-magnification relay (Optovar) which gives
overall image magnifications of 500X, 625X, 800X, or 1000X at the
oculars. The second path is focused onto a 64-element photodiode
line scanner by a 63mm photoprojection ocular; magnification at the
line array is such that it searches a 200-micra swath on the slide
at 3.18-micra resolution. This finder channel, in conjunction with
the finder logic in the image processor, locates objects which are
suitable for examination by the data acquisition channel. Based
on information derived by the finder, a 50 x 50 micra object field
is sampled in a 128 x 128 pixel format (0.4-micra pixel spacing) by
an instrument-grade television camera at the focal plane of another
63mm photoprojection ocular; at the exit pupil of this ocular are
mounted a galvanometer-driven mirror for centering objects located
by the finder and a semaphore carrying the 515nm and 580nm filters.
High-resolution data taken through these filters are used for focus
evaluation, electronic image centering, and object classification.

The finder line array operates at 925 lines per second, or
with an effective line spacing of 3.2 micra given the mechanical
slew speed of 3.2mm per second. As mentioned, the 200-micra line
length gives a resolution of 3.2 micra perpendicular to the stage
motion. The output of the line array is also averaged, to provide
a feedback signal used to stablize illumination intensity.

The controller for the data-acquisition camera effectively converts it into a special-purpose peripheral of the minicomputer, and
the operating software can adjust parameters of the controller to
maintain quality of the image data. Although the camera output is
taken at 0.4 micra pixel spacing and digitized in 6-bit, 128 x 128
format, software can command image access in either of two 64 x 64
formats. A "low-resolution" (0.8-micra pixel) image, obtained by
taking alternate pixels in alternate lines, is used to adjust focus,
to eliminate debris and to determine where a 25-micra square field
should be placed to electronically center the object. Once the 25
micra high-resolution field is located (its upper left-hand corner
may fall on any pixel within the upper left quadrant of the 128 x
128 image), a 64 x 64 pixel image in each color is scanned into
memory, for segmentation and feature extraction. Data acquisition
occurs within the camera frame period of 10.6 msec.

3.2   Optical Bench

The optics are mounted in a sturdy optical bench designed to
withstand the rigors of automated operation. Structurally, all
imaging optics are carried by a stiff, cast-aluminum bridge; the
finder and data-acquisition cameras are mounted on a rigid aluminum
casting to which the bridge is attached. The mechanical stage is
carried on a diaphragm flexure, free to be driven vertically by the
coarse-focussing mechanism but restrained from tilting by a central
linear bearing. Connecting the bearing shaft and the diaphragm is
a cup surrounding the condenser; this arrangement allows the condenser to remain stationary while the diaphragm is flexed by a cam
contacting the shaft below the bearing. Driven by a stepper motor,
the cam provides coarse-focussing steps of 0.8 micra over a range
of 500 micra. Fine-focussing steps of 0.12 micra, over a 7.5 micra
range, are generated by a piezo-electric element driving a cantilevered slide holder mounted on the mechanical stage. The stage
itself is a two-axis unit with ball ways, driven in each axis by
a precision lead-screw and stepper motor. Positioning resolution
is 3.2 micra over the 0.77cm by 2.0cm area through which the search
meander is normally programmed. And finally, the optical bench is
fitted with an autoloading mechanism, positioned so that removable
cassettes containing up to 14 slides can be readily mounted. In
summary, it will only be noted that the bench design can accept
any specimen mounted on beveled 1" x 3" microscope slides.

4. GOLAY IMAGE PROCESSOR

The programmable image processing hardware used in the diff3 analyzer is a parallel/serial implementation of a binary transform technique described by Golay (1969, 1977). Functional properties have been treated by Preston and Norgren (1972), and Preston (1971) has written of applications in feature extraction. In addition to providing programmable generation of the transforms themselves, the image processor does the histogramming and thresholding necessary for efficient processing of the images produced by the automated microscope (Steps 4 through 8 in Table 1). Separate logic circuits operating on the finder data also stop the mechanical stage (Step 3) when an object is found whose size exceeds a threshold on number of adjacent pixels darker than a reference density, but since this task is not central to image processing in the diff3 analyzer, it will not be considered further.

The basic aim during design of the image processor was to produce the most flexible hardware possible, in order not to restrict capabilities of the resulting instrumentation. Where engineering compromises were unavoidable, choices were based on characteristics needed to perform a 100-cell white cell (WBC) differential count, at the approximate rate of one count per minute. Three fundamental parameters were affected by these considerations. First, the image format was set at 64 x 64 pixels, experimental work having indicated that this format was sufficient for differentiating the normal white cells. Second, maximum image transformation time was set at 100 microseconds per image, so that 500-transform feature-extraction algorithms could be implemented at rates consistent with the specified differential rate. Third, four binary-image registers were included in the processor, so that the great majority of all feature extraction algorithms could be used without the necessity for retaining or reconstructing binary images. The organization of the resulting image processor is shown in Figure 3.

Image data, digitized into 6-bit grey levels, can be routed via either of two paths through the processor. In the upper one, it is compared to a threshold supplied as a programming input; resulting binary images are forwarded to the Golay LOgic PRocessor (GLOPR), for storage in one of the binary image registers discussed below. Alternatively, the 6-bit, grey-level image data may be presented to the histogrammer; if so, histogram data is recorded through one of the two 16-bit output registers directly into core memory of the controlling minicomputer via its I/O bus. The histogram table, containing the count of pixels in each of the 64 grey levels in the image data, may be analyzed in the minicomputer, to provide information such as the threshold values to be used in creating binary images. However, the most important output from the image processor is the count data supplied by the GLOPR. At the end of each binary

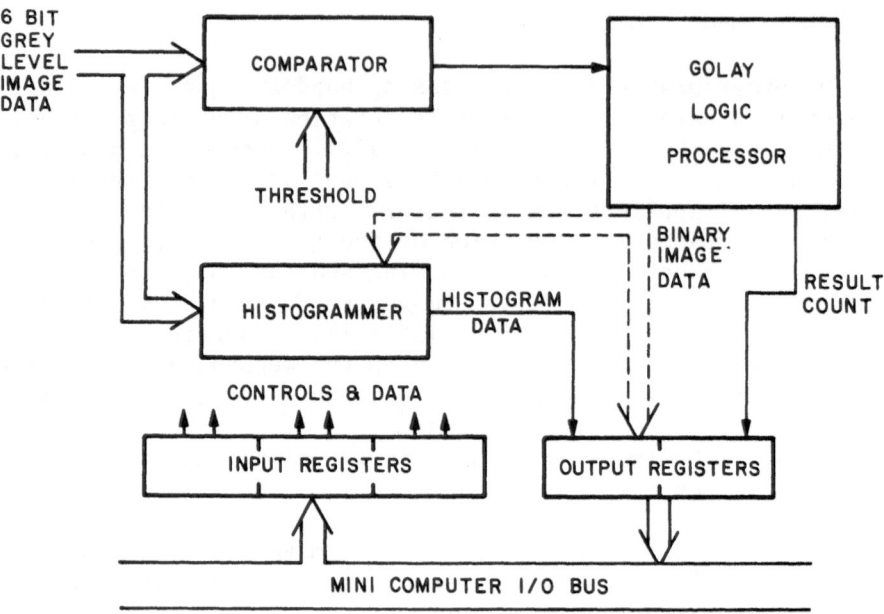

Fig. 3. Data paths in the Golay image processor.

image transformation, a count is made of the number of "1" bits resulting from that transform, and this count may be transferred on
command into minicomputer memory for storage as an image processing
result. In addition to such count data, the GLOPR may output binary
image data directly, for use in either of two ways. Binary images
may be routed to the histogrammer, where they serve as gates on the
photometric histogram; so used, they permit histogramming of areas
defined by the binary image, an extremely useful tool during image
segmentation. Binary images may also be routed via one of the two
output registers to minicomputer memory, if sufficient binary image
registers are not available to accomodate the full complexity of a
given algorithm; by this means the number of binary image registers
can be increased as necessary, using scratch arrays in core memory.
(Practically, it has been found that less than one per cent of image
processing tasks require this backup storage capability.) Control
of image processing functions by the minicomputer occurs by transfer
of control words and data through three 16-bit input registers driven by the I/O bus; in conjunction with similar programmability of
automated microscope functions, this provides an extremely flexible
array of feature-extraction capabilities with which to analyze
microscopical images, of whatever origin.

    The source of the 6-bit grey-level image data is the dedicated
memory in which the digitized output of the data-acquisition camera

is held. This memory, implemented with shift registers, is external
to the minicomputer and is organized into four 4096 (64 x 64) pixel
regions, into which either 0.8 or 0.4 -micra images may be loaded.

4.1 Golay Logic Processor (GLOPR)

The GLOPR contains the majority of the parallelism within the
diff3 image processor. As illustrated in Figure 4, the primary
parallel data path is a 16-bit wide bus which couples the four binary image registers, each holding 4096 points, and the transform
generator. The latter can accept either one or two binary images
from the registers, implement a specified transform operation, and
return the transformed image to any of the four binary registers.
In addition, a count of the "1" bits in the output image is made at
the completion of each transform. The transformation process itself is controlled by the five functions indicated in the figure.
This control information is passed as three 16-bit words, for com-
patability with 16-bit computers, and consists of the following:

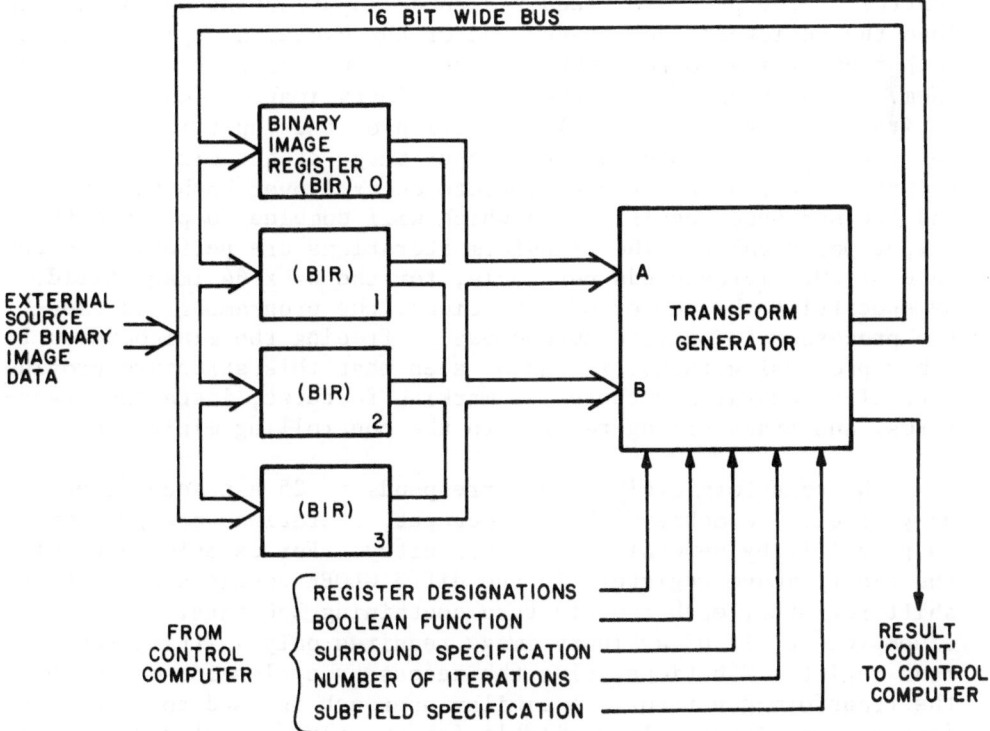

Fig. 4. Organization of the Golay logic processor.

1. Register designations. One or two source registers, plus a destination register, must be specified; this requires 6 bits of control data.

2. Boolean function specification. This provides the capability for the control computer to form all possible combinations of data from Sources A and B with the surrounds specified by a selected Golay function. 8 bits of control data are used.

3. Surround specifications. This defines the Golay function being sought and requires 14 bits, one for each of the surrounds designated by Golay.

4. Number of iterations which a given transform is to be performed. In the configuration used in the diff3 analyzer, 6 bits are devoted to this specification. Thus, from 1 to 63 iterations may be requested; the 64th value causes iteration until the image stabilizes (i.e., output = input).

5. Subfield specification. In the diff3 analyzer only subfields of one (the complete image) or subfields of three are implemented; thus, one bit suffices.

Control of the processor requires three input/output operations to load the registers, and at the end of the cycle, a separate input/output operation to read the processed result back from any one of them. Each iteration produces a new binary image which is (at all of its 3969 interior points) a function of the input image(s) at the corresponding point and the input image at the six surrounding points. The programmer has complete control over both the function and the surround specification which will combine to produce the output point value. The transform iterations are performed at the rate of 100 microseconds per cycle, for the 64 x 64 image field. By specifying the number of iterations, the programmer can cause the processor to operate autonomously, freeing the minicomputer for other processing tasks. It can be seen that this structure provides a totally general structure for marking features, iterating transforms, and transferring results to the controlling minicomputer.

The transform cycle time corresponds to 25 nanoseconds per binary pixel. To achieve this throughput, operations are performed in parallel, by several sets of circuitry. For example, each of the binary image registers in the diff3 GLOPR consists of a 16-rank shift register, each rank in turn containing 256 bits. Therefore, a complete cycle of a binary image requires only shifting the 16-rank register 256 times, plus the necessary cycles for delay through the transform generator. A similar approach is used to insure that data are simultaneously available for any given pixel and its surround of six neighboring points.

Shown below in Figure 5, this latter hardware consists of a second 16-rank shift register, but one containing only 11 elements. Data entering the transform processor from one of the binary registers is passed into this 11-stage, 16-rank register at a clock rate of 2.5 MHz. During any given clock period this register contains two complete 64-pixel lines, plus 48 pixels from a third line. Information is tapped off the register to provide signals to 8 sets of circuitry for generation of Golay transforms. Since it requires 16 pixel cycles to equal one cycle period on the external bus, each of the 8 transform circuits is time-shared to process two pixels and their corresponding surrounds during a single cycle on the external bus. The data path is through the transform circuitry, the pixels merely overflowing the 16-rank, 11-element register as new ones enter.

Each of the 8 transform generation circuits consists of an arrangement such as that shown in Figure 6. A read-only memory (ROM) containing 64 words of 14 bits each is used as the basic functional device. The pattern written into the ROM is such that each word comprises 13 "0" bits and a single "1" bit, the position of which identifies the Golay index of the surround pattern addressing that particular word. The six bits taken from a surround, contained in the 16-rank shift register, are applied as address bits to the ROM. The resulting output word specifies, by the location of its "1" bit, the surround pattern present at that point in the image. As shown in Figure 6, the 14 bits of the ROM output are each applied to an input of a two-input AND gate; the other inputs for this array of AND gates come from the Surround Specification supplied as programming information to the image processor. The output from the

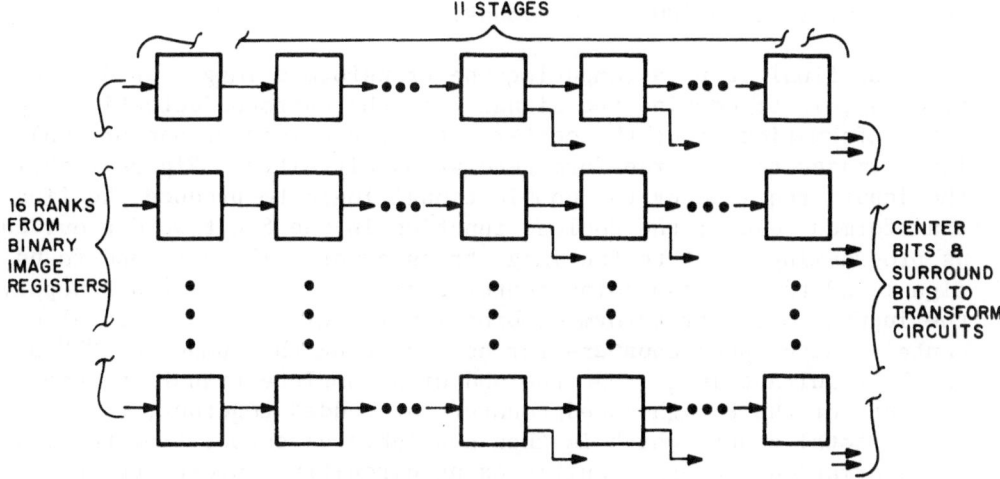

Fig. 5. Structure used to synchronize surround data.

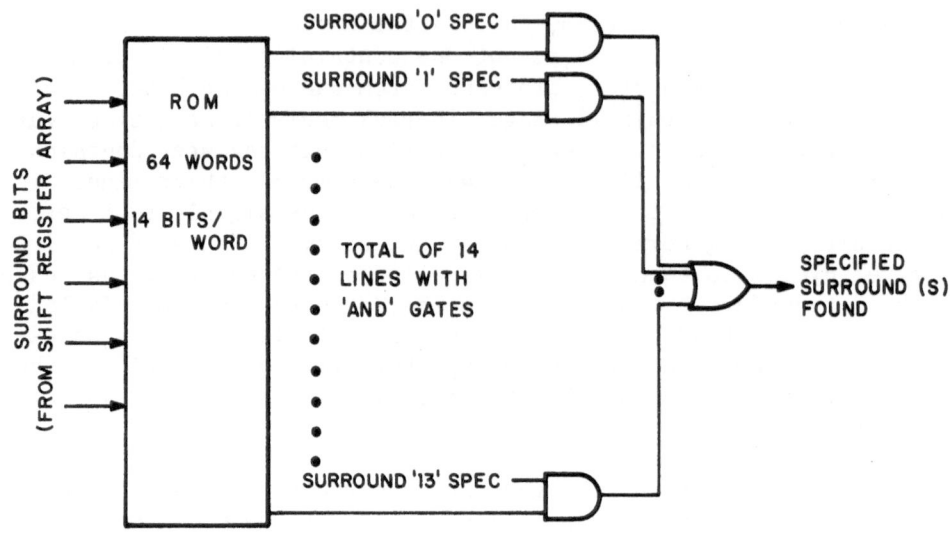

Fig. 6. Transform generation circuits used in GLOPR.

individual AND gates is logically "ORed" to generate a signal indicating when the specified surround(s) have been found. Thus, the transform can be generated on the basis of any individual surround or of any desirable combination of surrounds. As has been noted, this circuitry is replicated 8 times, to produce the parallelism required by speed considerations. Consequently, a basic cycle time of 200 nanoseconds for each circuit provides a total transform rate of one every 25 nanoseconds, and a complete 4096-bit transform can be accomplished in 100 microseconds.

The final step in computing the transformed image, as defined by Golay, is to combine the signal from the surround detection logic with information about the center bits of the (either one or two) input images to the transform generation circuitry. Figure 7 shows the inputs required by the combinational logic to produce the final transformed result; the Boolean function is the 8-bit word provided as programming input to the image transformer. The surround-found signal and the corresponding center bits from images A and B supply the inputs to the transform combinational logic. This logic also contains high-speed counters for accumulating the number of "1" bits in the resultant image. At the end of a complete transform, the contents of the 8 high-speed counters are added together, to produce a total count, which is then available as output from the image transformation process. Multiplexing circuitry converts the 8-bit parallel processing stream into 16-bit, bus-compatible form.

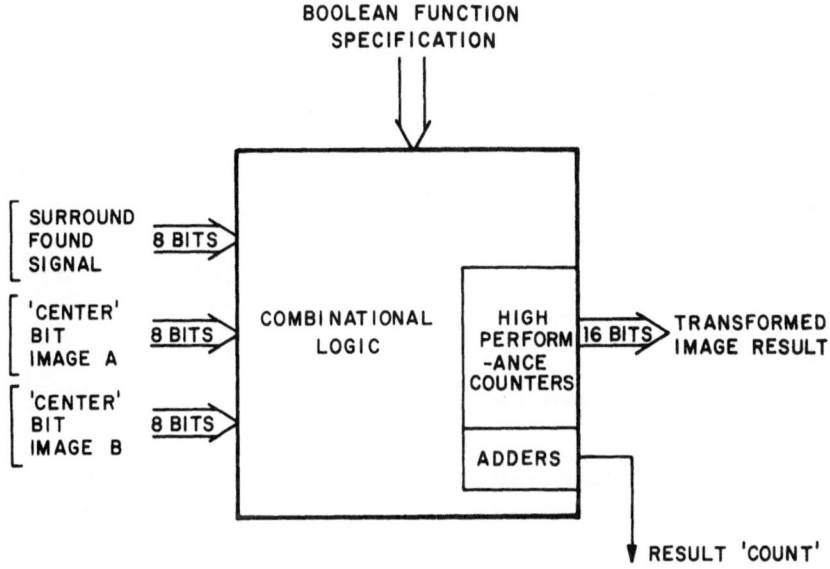

Fig. 7.   Combinational logic circuits used in GLOPR.

4.2 Application of GLOPR

As an example of the use of GLOPR circuitry, consider the task of segmenting into its components an image field scanned from a blood smear; such fields commonly contain a background, white cells (WBC) consisting of cytoplasm surrounding a nucleus, red cells (RBC) which contain a central pallor, and platelets. Segmentation is accomplished by a complex program which uses the histogrammer, in conjunction with the Golay image transforms. First, the image histogram is analyzed, to detect peaks and valleys and to select suitable thresholds for segmenting the scene. Using the thresholds, the comparator generates binary images, which are stored in the binary registers for transformation. The GLOPR performs "cleanup" steps and generates a mask (or cartoon) of the scene, as illustrated in Figure 8; the mask is used in turn to generate a gated histogram containing only data from those points set to "1" in the mask. The cleanup operation is illustrated in Figure 9; objects touching the border of the white cell are removed from the stored image. This is done by trimming all objects a certain number of times, until images of touching objects separate; expanding the objects back, but without letting them reform contact; and selecting only the object containing a nucleus. Without the GLOPR, it would be difficult

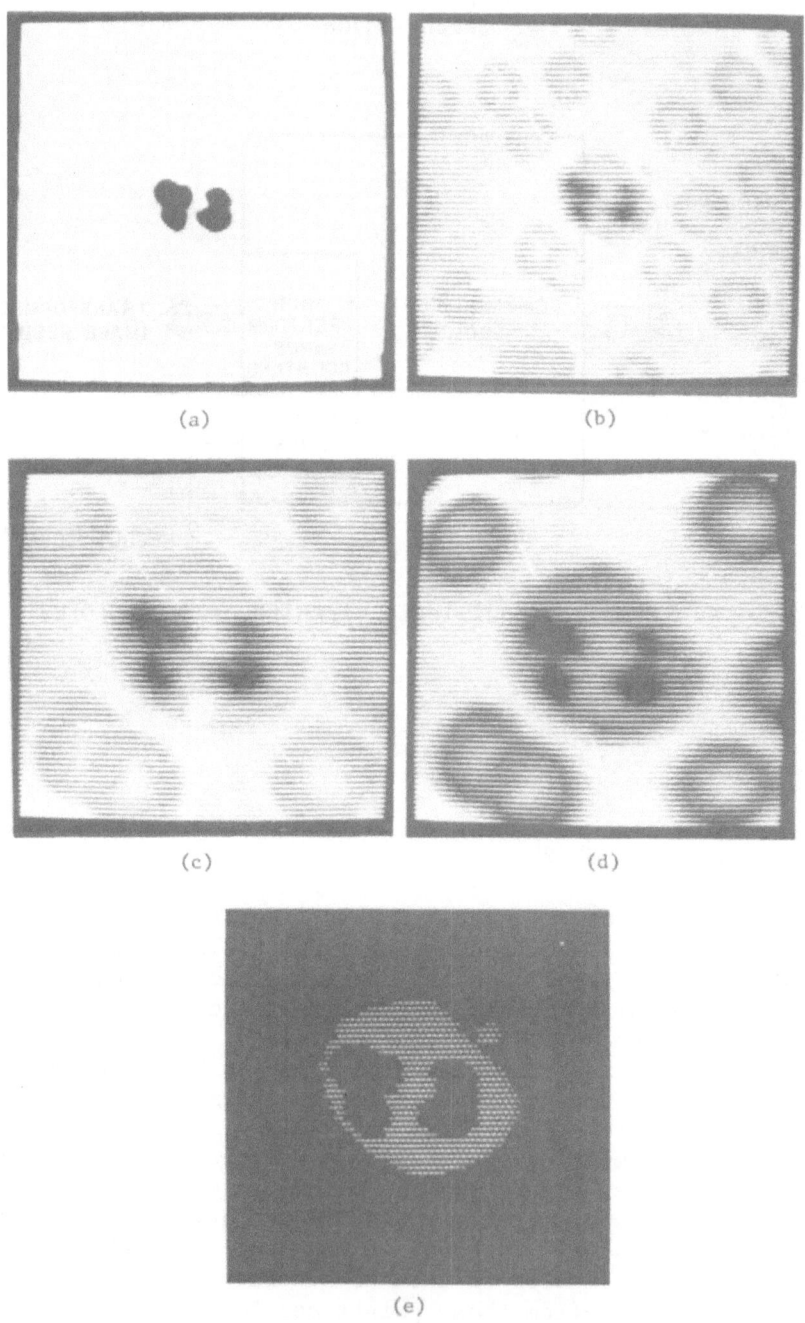

Fig. 8. Steps in cell acquisition. (a) Video image. (b) Low-resolution red image. (c) High-resolution red image. (d) High-resolution green image. (e) Cartoon of cell cytoplasm.

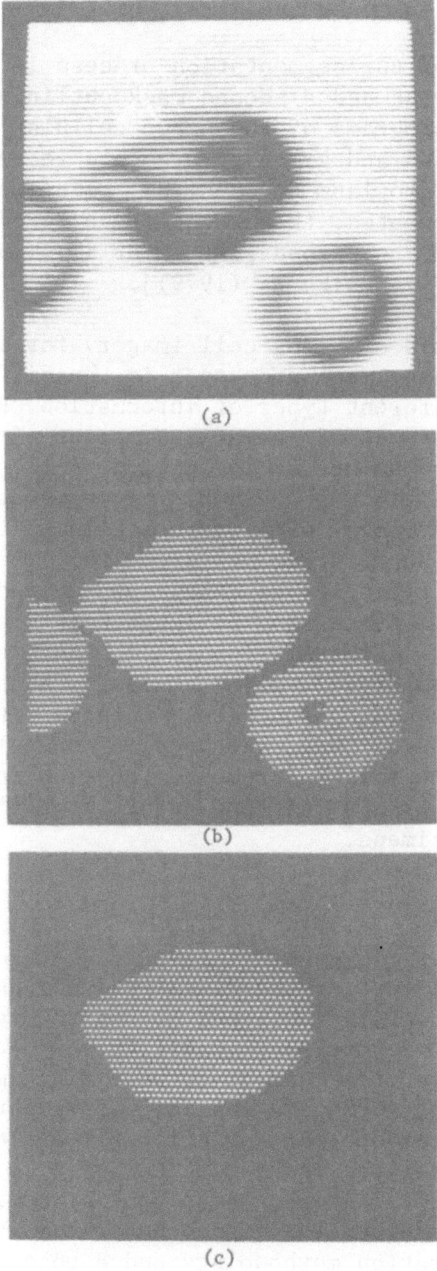

Fig. 9. Example of image "cleanup". (a) Red image of a leukocyte with one touching and one non-touching red cell. (b) Cartoon of cells prior to cleanup. (c) Cartoon of leukocyte after
red cell has been trimmed off.

to separate platelets touching white cells, or red cells touching
nucleated red cells, on the basis of color alone.

The product of the segmentation process is a mask showing the
points in the nucleus and a second mask outlining the cytoplasmic
points. Gated histograms of these two cellular components are constructed in both red and green colors, and features such as mean
densities of the cytoplasm and nucleus, color, contrast and other
descriptors are computed, the use of which for differentiation of
leukocytes has been described [Ingram et al. (1968), Brenner et al.
(1974), Green (1979), Kulkarni (1979)].

In the analysis of blood-cell imagery for the purpose of cell
differentiation, the image processor in the diff3 analyzer is used
to provide two different types of information. Result counts taken
after various sequences of operations measure features related to
the shape, size and number of the several cell classes and constituents. In addition, the binary images produced by the image
processor serve as masks defining the cellular constituents for the
purpose of making photometric measurements. The commonalty of the
two informational types is that they both depend on spatially related data, and the processing of image data according to its spatial relationships is the distinguishing feature of processors such
as that used in the diff3 analyzer. It is the speed of the histogramming and thresholding functions, coupled with the programmable
parallel architecture, that makes real-time analysis of complex
images practical in this instrument; these characteristics are of
equal value in other applications requiring analytic processing of
slide-mounted specimens.

5.   SUMMARY

Internal structure of the diff3 analyzer has been described.
Configured as a specially designed automated microscope and image
processor, both of general applicability and operating as special-
purpose peripherals of a general-purpose minicomputer, the analyzer
has been applied to automated analysis of Wright's stained blood
smears. However, except for the color filters and objective used
with the microscope, the instrument hardware is general purpose;
characteristics defining it as a hematology device are contained in
the programs loaded into the minicomputer memory. Thus, with suitable sample preparation methodology and a possible change of color
filters and objective, application of the analyzer's capabilities
to any of the many samples capable of being prepared as 1" x 3"
slide mounts only requires development of suitable software. Thus,
power of real-time parallel processing techniques may potentially
be applied to a broad range of biomedically important specimens.

6.   REFERENCES

Brenner, J. F., Gelsema, E. S., Necheles, P. W., Neurath, P. W.,
    Selles, W. D., and Vastola, E., "Automated Classification of
    Normal and Abnormal Leukocytes," J. Histochem. Cytochem.
    22:697 (1974).

Gilmer, P. R., "The Coulter Perkin-Elmer diff3 Automated Counting
    System," in Differential Leukocyte Counting (Koepke, J. A.,
    ed.) College of American Pathologists, p.135 (1978).

Green, J. E., "A Practical Application of Computer Pattern Recog-
    nition Research - The Abbott ADC-500 Differential Counter,"
    J. Histochem. Cytochem. 27:160 (1979).

Golay, M. J. E., "Hexagonal Parallel Pattern Transformations,"
    IEEE Trans. Comput. C-18:733 (1969).

Golay, M. J. E., "Analysis of Images," U.S. Patent 4,060,713,
    November 29 (1977).

Ingram, M., Norgren, P. E., and Preston, K., Jr., "Automatic Dif-
    ferentiation of White Blood Cells," in Image Processing in
    Biological Science (Ramsey, D. M., ed.) Univ. of Calif. Press
    p.97 (1968).

Kulkarni, A. V., "Effectiveness of Feature Groups for Automated
    Pairwise Leukocyte Class Discrimination," J. Histochem.
    Cytochem. 27:210 (1979).

Maher, F. W., and Pirc, V. V., "Method for Blood Film Preparation,"
    U.S. Patent 4,037,003, July 19 (1977a).

Maher, F. W., and Pirc, V. V., "Apparatus for Blood Film Prepa-
    ration," U.S. Patent 4,016,828, April 12 (1977b).

Maher, F. W., and Pirc, V. V., "Diluting and Dispensing Probe for
    Blood Sample Preparation," U.S. Patent 4,108,608, August 22
    (1978).

Megla, G. K., "The LARC Automatic White Blood Cell Analyzer," Acta
    Cytol. 17:3 (1973).

Miller, M. N., "Design and Clinical Results of Hematrak: an Auto-
    mated Differential Counter," IEEE Trans. Biomed. Eng.

Norgren, P. E., Kulkarni, A. V., and Graham, M. D., "Leukocyte Image Analysis in the diff3 System," submitted to Pattern Recog. (1980).

Nosanchuk, J. S., Dawes, P., Kelly, A., and Heckler, C., "An Automated Blood Smear Analysis System," Am. J. Clin. Pathol. 73:165 (1980).

Nourbakhsh, M., Atwood, J. G., Raccio, J., and Seligson, D., "An Evaluation of Blood Smears Made by a New Method Using a Spinner and Diluted Blood," Am. J. Clin. Pathol. 70:885 (1978).

Preston, K., Jr., "Feature Extraction by Golay Hexagonal Pattern Transforms," IEEE Trans. Comput. C-20:1007 (1971).

Preston, K., Jr., and Norgren, P. E., "Interactive Image Processor Speeds Pattern Recognition by Computer," Electronics 45:89 (1972).

Preston, K., Jr., Duff, M. J. B., Levialdi, S., Norgren, P. E., and Toriwaki, J.-I., "Basics of Cellular Logic with Some Applications in Medical Image Processing," Proc. IEEE 67:826 (1979).

Preston, K., Jr., "Automation of the Analysis of Cell Images," Anal. Quant. Cytol. 2:1 (1980).

Schoentag, R. A., and Pedersen, J. T., "Evaluation of an Automated Blood Smear Analyzer," Am. J. Clin. Pathol. 71:685 (1979).

COMPUTER-AIDED TISSUE STEREOLOGY

K. Paton

Division of Medical Computing, Clinical Research Center

Watford Road, Harrow, GREAT BRITAIN

1.   INTRODUCTION

Stereology (Underwood, 1970; Miles and Davy, 1976) is the science of inferring multi-dimensional spatial structure from lower-dimensional data which are usually in the form of sections or projections. Such partial information (the lower-dimensional data) is called "*stereological data*" and has traditionally been gathered by counting the intersections of the structures of interest with a suitable test grid. We describe three problems to which stereological methods may be applied and show that in all three cases the data may be easily gathered by minor variations of a single method of computer-aided image analysis.

2.   ANALYSIS OF MUSCLE BIOPSIES

In the study of neuromuscular disease (Adams et al., 1968) it is important to assess the spatial patterns formed by different types of fibers found in sections of muscle biopsy. It is tradi-tionally assumed that each fiber is a circular cylinder and that compression together distorts individual fibers in such a way that they fill space. In a typical cross-section of muscle, such as Figure 1, each fiber corresponds to a more-or-less polygonal patch: the light ones are known as type I; the dark ones, type II.

Important questions therefore include, "Do the type I fibers cluster together or are they uniformly dispersed?" and "How big are the fibers?" as well as "How many fibers are there of each type?" The second question is a stereological one.

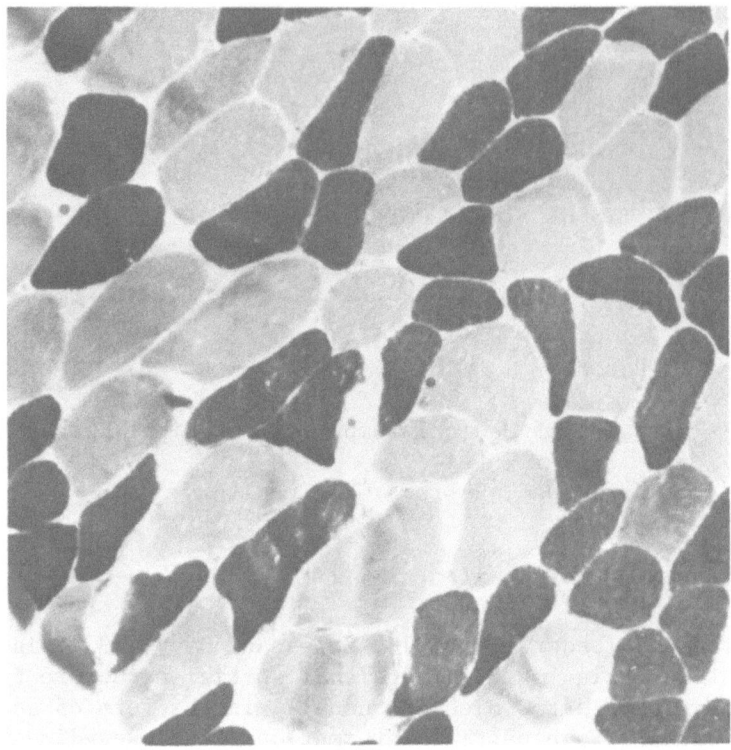

Fig. 1.   Section of muscle biopsy.

     One traditional method of estimating fiber size relies on the
observation that, if a circular cylinder of diameter d be cut by
any plane, then the plane section is an ellipse of minor axis d.
Hence fiber size may be estimated by treating each fiber section
as an ellipse and measuring its minor axis.   In the manual method
in use in the Department of Pathology, Northwick Park Hospital,
the observer covers a photograph of the muscle biopsy section with
a sheet of translucent paper and records thereon the longest chord
perpendicular to the longest diameter of each fiber section.   When
all fiber sections have been so treated the record on the overlay
consists of one *width vector* for each fiber section, the length of
this vector being the estimated width of the fiber.   Although this
process is subjective, it is not known whether the effect is
important.   When the major and minor axes of the patch are very
different it is comparatively easy to choose the "right" direction;
as the major and minor axes tend to equality it becomes harder to
choose the "right" direction, but the error committed in choosing
a "wrong" direction tends to zero.   The time taken to obtain the
data for the width distribution for 200 fibers by manual methods is
is about 90 minutes.

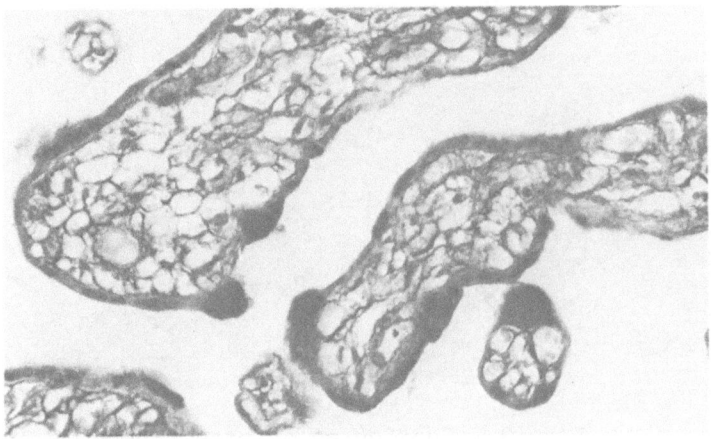

Fig. 2. Section of human placenta.

3.   ANALYSIS OF THE PLACENTA

     In the study of the placenta it is important to relate its
physical components to the capacity for transport of gases and
nutrients between these components. In the section of the
placenta shown in Figure 2 the darker components are the villi,
belonging to the foetus, and the light background is inter-villous
space, belonging to the maternal circulation; some of the light
discs within the villi are capillaries. Transport takes place
between the intervillous space and the villous capillaries. It
is of interest to estimate the volume fraction of these components.

     Laga et al. (1973) describe methods for estimating volume
fractions by computer-aided point-counting. Without such aids the
task is tedious but not particularly time-consuming. To obtain an
accuracy of 5% in the volume fraction of the villi it is necessary
to count some 200 points on the villi. Counting at a rate of
4 points per second, the human observer would need only about 1
minute to obtain the data on the volume fraction of the villi for
one specimen.

4.   ANALYSIS OF ADIPOSE TISSUE

     In the study of obesity (Ashwell et al., 1976) it is important
to determine whether more fat is put into existing fat cells or
whether new fat cells are formed. Fat cells are more or less
spherical objects in three-dimensional space; in a typical slide
of a cross-section of adipose tissue (Figure 3), each cell is

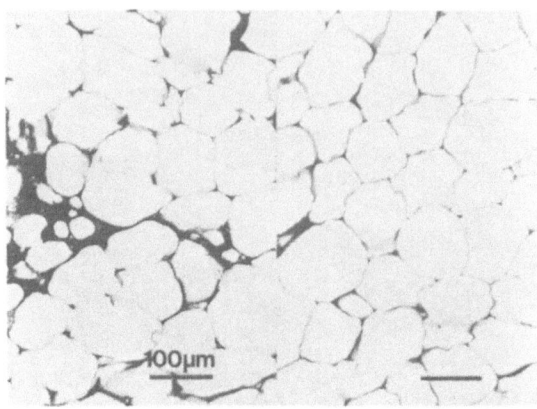

Fig. 3。 Section of adipose tissue.

represented by a more or less circular white patch delineated by
a thin grey or black boundary. A patch of radius R may be produced
by an equatorial section of a sphere of radius R, as a near-polar
section of a sphere of radius much larger than R, etc. It is
desired to find the size distribution of the spheres from the size
distribution of their plane sections. If the spheres are well
separated in space then the method of Schwarz-Saltikoff (see DeHoff
and Rhines, 1968) would allow the size distribution of the spheres
to be deduced from the size distribution of their plane sections.
However, if the fat cells are more-or-less space-filling, as in
Figure 3, this approach is not valid.

     To obtain the desired information by manual methods 40 serial
sections must be counted; to obtain an accuracy of 5% it is necessary to count some 200 points on each cell section. Counting at a
rate of 4 points per second, the human observer would need some 33
minutes to obtain the data.

5.   COMPUTER-AIDED DATA CAPTURE

     The data capture for each of the three problems described above
consists of two activities: (1) *segmentation* of an image into its
components and (2) *measurement* of properties on the components thus
found. Methods for data capture may be classified thus:

Method	Segmentation	Measurement
Manual	Observer	Observer
Computer-aided	Observer	Computer
Automatic	Computer	Computer

Comparatively speaking, the human observer is good at segmentation but poor at measurement, whereas the computer is poor at segmentation but good at measurement. This poses two types of challenge: (1) to devise methods of automatic segmentation which rival those of the human observer, and (2) to combine the skills of observer and computer in order to capture the data most effectively. The program for automatic segmentation of muscle fibres reported by Eccles et al. (1977) is a response to this type of challenge.

The challenge offered at the Clinical Research Center is of type (2). A software system was therefore written for the Joyce-Loebl Magiscan (Taylor and Dixon, 1976) in which the observer segments the field of view with the light pen and the computer measures appropriate properties of the regions thus defined.

For fat cells or the placenta the property of interest is simply area. For muscle we must derive the minor axis of the nearest ellipse having the same moments as the fiber cross-section. This is derived as follows. Let m_{ij} be the $(i, j)^{th}$ central moment of the region. Let $\lambda_2 \geqslant \lambda_1 \geqslant 0$ be the eigenvalues of the matrix

$$M = \begin{vmatrix} m_{20} & m_{11} \\ m_{11} & m_{02} \end{vmatrix} \tag{1}$$

If the region were an elliptical lamina of unit mass and with axes 2a and 2b ($a \geqslant b$), then these eigenvalues would be $a^2/4$ and $b^2/4$. Hence the minor axis of the region may sensibly be obtained by equating $b^2/4$ with λ_1 and the width of the fiber section defined as $4\sqrt{\lambda_1}$.

Tables 1 and 2 summarize the results of two experiments to compare the effectiveness of data capture by the manual method and the computer-aided methods described above. In each case a photograph was prepared in which the objects of interest were indicated in such a way that the same objects were repeatedly measured. Each observer used each method on each of two days. Since the true value for the measurement is unknown, the absolute accuracy of the process is unavailable; instead the quality of the data capture may be assessed in terms of reproducibility, viz. the standard deviation of the between-day or between-observer differences.

Let x_{fmjt} denote the reading made by observer j on object f using method m on occasion t. Let the number of objects be F. Then the difference d_{fmj}, mean difference $\bar{d}_{jm}$, and variance s^2_{jm} of difference are defined respectively as:

$$d_{fjm} = x_{fjm1} - x_{fjm2}$$

$$\bar{d}_{jm} = \sum d_{fjm}/F \tag{2}$$

$$s_{jm}^2 = \left(\sum d_{fjm}^2 - F\bar{d}_{jm}^2\right)/(F-1)$$

Further, let y_{fjm} denote the average of the two readings made by observer j on object f using method m. Then the difference d_{fm}, mean difference $\bar{d}_m$, and variance s_m^2 of difference are defined respectively as

$$d_{fm} = y_{f1m} - y_{f2m}$$

$$\bar{d}_m = \sum d_{fm}/F \tag{3}$$

$$s_m^2 = \left(\sum d_{fm}^2 - F\bar{d}_m^2\right)/(F-1)$$

Tables 1 and 2 refer to measurements of the "width" of muscle fibers and "diameter" of fat cells, respectively. In each case the designated table "a" is based on between-day differences for each observer; "b" on between-observer differences. The tables show that a substantial increase in speed can be obtained without appreciable loss in reproducibility

Table 1 - Analysis of 50 Muscle Fibers

(a) Between-Day Differences				
Method	Observer	Mean (microns)	S.D. (microns)	Time (minutes)
m	j	$\bar{d}_{jm}$	s_{jm}	T_{jm}
manual	1	−0.08	3.28	26
	2	0.24	2.96	21
computer-aided	1	−0.18	4.98	4.5
	2	3.18	3.00	4

Table 1 - (continued)*

(b) Between Observer Differences			
Method	Mean (microns)	S.D. (microns)	Time (minutes)
m	$\bar{d}_m$	s_m	T_m
manual	2.48	2.41	23.5
computer-aided	1.66	3.63	4.3

*The average value for a muscle fiber is 60 microns.

Table 2 - Analysis of 33 Fat Cells*

(a) Between-Day Differences				
Method	Observer	Mean (microns)	S.D. (microns)	Time (minutes)
m	j	$\bar{d}_{jm}$	s_{jm}	T_{jm}
manual	1	-2.09	5.66	10.5
	2	4.38	8.10	13.5
computer-aided	1	1.12	4.13	5.5
	2	0.33	2.95	7.0
(b) Between-Observer Differences				
Method		Mean (microns)	S.D. (microns)	Time (minutes)
m		$\bar{d}_m$	s_m	T_m
manual		2.67	4.56	12.0
computer-aided		-2.70	3.64	6.3

*The average value for a fat cell is 88 microns.

6.   CHOICE OF METHOD

We may attempt to capture stereological data by fully automatic methods (automatic segmentation and automatic measurement),
computer-aided methods (such as the one described above) or traditional methods (such as point counting). The flow chart below
shows a decision tree for selecting an appropriate method. Group A
and Group B are the terms introduced by Weibel (1967). In Group A
the specimen contains convex solids whose number and sizes are required; in Group B the components weave tortuously through the
specimen and it is required to find the volume fraction of each
component.

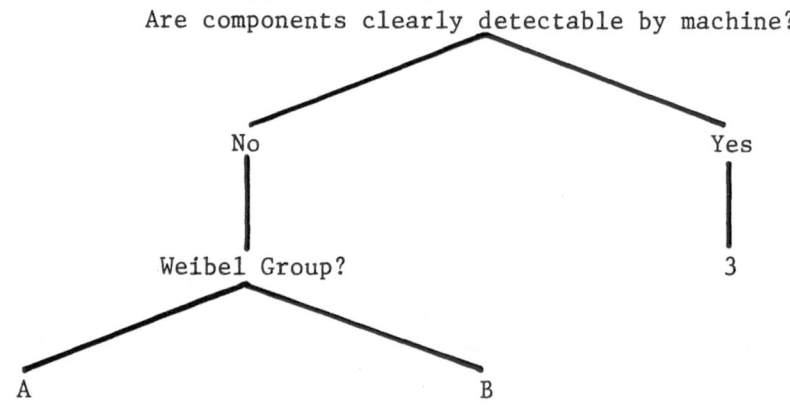

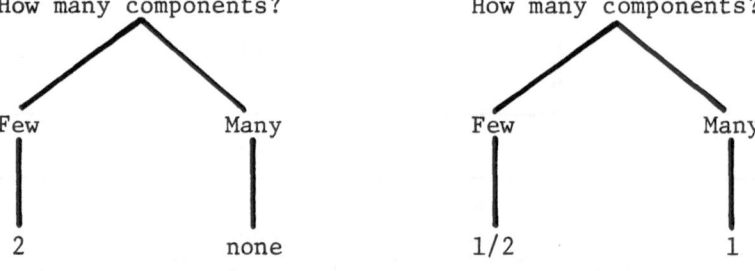

Legend: 1 - point counting

           2 - computer aided

           3 - automatic

Let Method 1 denote point-counting, Method 2 denote manual segmentation followed by automatic measurement and Method 3 denote automatic segmentation by grey-level thresholding followed by automatic measurement. When components are not clearly detectable by machine, the choice lies between traditional and computer-aided methods. The choice in Group B depends on the number of components. If there are few components there is little to choose between Method 1 and Method 2; if there are many components Method 1 is the more effective. This is because the time to count the points is independent of the number of components whereas the time for manual segmentation increases with the number of components.

In Group A, Method 1 must be discarded since many points, say 200, are required for each component. When the number of components is small, Method 2 is satisfactory; when the number of components is large, none of the three methods is satisfactory.

7. ACKNOWLEDGEMENTS

The advice of Dr. M. Ashwell, Dr. F. E. Hytten and Dr. G. Slavin is greatly appreciated.

8. REFERENCES

Adams, R. D., Coers, C., and Walton, J. N., "Report of a Sub-Committee on the Quantitation of Muscle Biopsy Findings," J. Neuro. Sci. 6:179-188 (1968).

Ashwell, M., Priest, P., Bondoux, M., Sowter, C., and McPherson, C. K., "Human Fat Cell Sizing--A Quick Simple Method," J. Lipid Res. 17:190-194 (1976).

DeHoff, R. T., and Rhines, F. W., Quantitative Microscopy, New York, McGraw-Hill (1968).

Eccles, M. J., McQueen, M. P. C., and Rosen, D., "Analysis of the Digitized Boundary of Planar Objects," Pattern Recog. 9:31-41 (1977).

Laga, E. L., Driscoll, S. G., and Munro, H. N., "Quantitative Studies of Human Placenta," Biol. Neonate 23:231-259 (1973).

Miles, R. E., and Davy, P. J., "Precise and General Conditions for the Validity of a Comprehensive Set of Stereological Fundamental Formulae," J. Micros. 107:211-226 (1976).

Taylor, C. J., and Dixon, R. N., _Quantitative Image Analysis Using Structural Methods_ (Hay, G. A., ed.), New York, Wiley (1976).

Underwood, E. E., _Quantitative Stereology_, Reading, Massachusetts, Addison-Wesley (1970).

Weibel, E. R., Quantitative Methods in Morphology, Berlin, Germany, Springer (1967).

INTERACTIVE SYSTEM FOR MEDICAL IMAGE PROCESSING

K. Preston, Jr.

Carnegie-Mellon University

Pittsburgh, PA   15213   USA

1.   INTRODUCTION

A joint program in high-speed, interactive medical image processing has been pursued during the past 4 years between the Department of Electrical Engineering of Carnegie-Mellon University and
the Department of Radiation Health, School of Public Health, University of Pittsburgh. An image processing facility called BIPU
(Biomedical Image Processing Unit) has been constructed around the
Perkin-Elmer 7/32 computer at Presbyterian Hospital within the
University of Pittsburgh Health Center. This system is interfaced
to various peripheral equipment through two additional Perkin-Elmer
6/16 computers and is also interfaced via both selector channels
and the multiplexor bus to a large high-speed disc, several CRT
terminals (both on-site and remote), line-printer, image printer-
plotter, etc. Projects using this facility at the present time in-
volved studies in computed tomography, ultrasound imaging and signal
analysis, dental-facial image processing, and the analysis of images
generated by the television microscope of tissue from various organ
systems.

Interaction is provided for the individual user via the standard CRT terminal backed up by means of several gray scale television displays and a full-color Ramtek display. Direct connections
to different experimental areas in the School of Public Health and
the Medical School are furnished. In particular real-time signal
processing in therapy planning, electronic radiography, and animal
toxicity studies are furnished.

An image processing language has been developed called
"SUPRPIC." Since FORTRAN provides the facility for processing
arrays, i.e., images, it is frequently used as the vehicle for
coding image processing software systems. SUPRPIC is coded in
FORTRAN with emphasis on a structure which permits the user to
select image processing commands from a menu. The goal has been
to execute any menu command in approximately 1 second. Thus
SUPRPIC is coded in such a way that all images, which are stored
on disk due to the usual limitations of core, are appropriately
buffered for high speed transfer to and from the computer core.
Since even the best FORTRAN compilers are frequently inadequate to
provide the optimized assembly language needed for rapid number-
array manipulation, all inner loops in SUPRPIC are coded in assembly language. The Perkin-Elmer FORTRAN 5 compiler is particularly
attractive in that it permits mixed FORTRAN and assembly-level
language coding. Furthermore, when the best assembly-level language is inadequate in speed, it is possible to revert to hardware
implementation of subroutines which cause significant bottlenecks.
The Perkin-Elmer computer system provides the facility for incorporating user-design, special-purpose peripherals for high-speed subroutine implementation. Such peripherals are implemented in a
standard Universal Logic Interface (ULI) board and may be accessed
either via the selector channels or the multiplexor bus.

2.   USING THE SUPRPIC INTERACTIVE IMAGE PROCESSING SYSTEM

SUPRPIC is an outgrowth of the language called "GLOL" with
certain major differences. GLOL is a symbolic language as described
originally by Preston (1972). It is based on the parallel image
transform of Golay (1971). In order to equate two images in GLOL,
the user enters A=B. When using SUPRPIC the user enters the number
string "0409010902." As is described below, each pair of digits in
this string provide pointers to either the command, image locations
in memory, or to command parameters. In the example given, the
numeric 04 points to the COPY command, the numerics 0901, to the
first logical image buffer in logical unit 9, 0902 to the second
logical image unit buffer in logical unit 9. The result is to
cause the system to copy the contents found in SUPRIC location 0901
on the disk into location 0902 on the disk. Execution time is
approximately 0.5 second on the Perkin-Elmer 7/32 (0.2 second on
the Perkin-Elmer 8/32) when operating upon a 512x512 image. The
reason for requiring the user to type numeric entries entirely is
to permit use of only one row of keys on the keyboard without
employing the shift key. This permits extremely fast entry of
commands so as to be compatible with the fast execution times
available.

SUPRPIC is a collection of approximately 20 major subroutines coded by the author during the 1978-1979 interval and specifically designed to process large, multi-color images on a dedicated computer having instantaneous display. The Perkin-Elmer 7/32 computer was selected because of its advanced architecture, high speed/cost ratio, and powerful instruction set. The instruction set supports approximately 200 instructions capable of addressing bits, bytes, half-words, and full-words. The FORTRAN runtime library of this computer has language extension which may be used for list processing. It also supports bit and byte instructions capable of directly addressing up to 8 million locations. The machine also has the capability of addressing up to 255 ULIs which may be designed by the user as hardware implementations of frequently used subroutines. In the case of SUPRPIC this strategy permits the execution of cellular logic transforms in hardware with a speed increase of about 1000 over that possible using an assembly code produced by the FORTRAN compiler or of about 100 over manually optimized assembly code. This leads to an effective speed of about 50 million instructions per second on this general purpose 32-bit "minicomputer."

2.1 Cellular Logic for Image Processing

Cellular logic is a Boolean image algebra which is performed on arrays of logical numbers. Such arrays are produced by thresholding the arrays of real numbers which represent images. An example of the representation of an array of real numbers using several arrays of logical numbers is given in Figure 1. Each cellular logic operation (CLO) is the multivariate logical function which is used to transform the set of logical values which represent a logical image into a new set of logical values representing the results image. The present value of each picture element in the logical image along with the logical values of its directly adjacent picture elements are taken as the independent variables. The dependent variable is the logical number which constitutes the new logical value of each picture element. Sequences of CLOs may be written for the purpose of extracting rather complicated shape, size, context, and texture features from images.

In the Carnegie-Mellon SUPRPIC interactive image processing language the cellular logic command is called "AUGRED" (a contraction of "AUGMENT/REDUCE" as initially introduced by Preston and Norgren, 1973, 1979). The AUGRED subroutine computes the following quantities for the jth picture element and its 8 neighbors:

$$FAC_j = \sum X_{ij} \qquad\qquad CNUM_j = \sum |X_{ij} - X_{i+1,j}| \qquad (1)$$

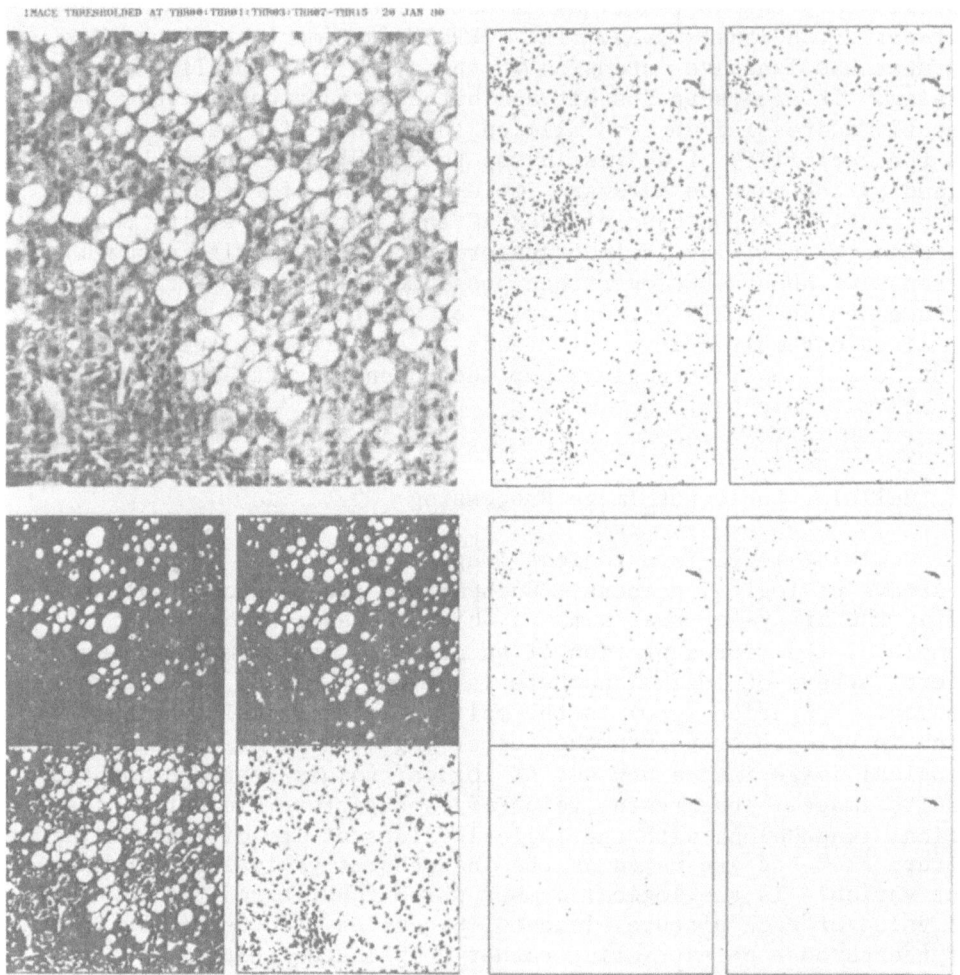

Fig. 1. A grayscale image may be represented by a set of logical images by thresholding. SUPRPIC uses 16 thresholds taken linearly between the 0.5% point and the 99.5% point of the probability density function. A grayscale image and logical images at thresholds 00,01,03,07,...,15 are shown above.

where X_{ij} ($i=1,2,\ldots8$) are the logical values of the 8 neighbors in the Cartesian tessellation, FAC_j is the number of logical 1s in the jth neighborhood and $CNUM_j$ is the number of 0-1 or 1-0 transitions in that neighborhood. (Note that $FAC_j=8$ indicates an interior point in a cluster of contiguous 1s; $FAC_j=0$, an exterior point; $CNUM_j=2$, an edge point; $CNUM_j=4$, a link in a chain of 1s; etc.) The value of the dependent variable is calculated by first computing the values of two intermediate variables A_j and B_j which are given by:

$$A_j = \begin{cases} 0 & \text{if } FAC_j \leq FAC \\ 1 & \text{otherwise} \end{cases}$$

$$B_j = \begin{cases} 0 & \text{if } CNUM_j < CNUM \\ 1 & \text{otherwise} \end{cases}$$

where FAC and CNUM are parameters of the AUGRED command. Finally, the logical value of the dependent variable is given by the Boolean algebraic expression:

$$X'_j = X_j(A_j+B_j) \tag{2}$$

where X'_j is the value of the dependent variable and X_j is the logical value of the independent variable representing the jth picture element.

The AUGRED command has three other arguments: CYC, BORD, MODE. The quantity CYC indicates the number of interations which are to be performed over the NxN array; BORD, the logical value of picture elements on the border of the array; MODE, the subfield sequence. When using the Cartesian tessellation, four subfields are utilized. These subfields divide the NxN array into four disjoint sets in such a way that a picture element belonging to one of these sets has a neighborhood whose elements reside in the other three sets as more fully discussed in Preston et al. (1979). When MODE=0, subfields are not utilized; MODE=k ($k=1,2,\ldots,4$), only the kth subfield is utilized in executing the CLO; MODE=5, four iterations using the subfield sequence 1-3-4-2; MODE=6, four iterations with the subfield sequence 1-3-4-2 but with the retention of residues (a picture element having the logical value 1 all of whose neighbors have the logical value 0).

2.2   The Command Structure of SUPRPIC

     The entire SUPRPIC command menu is presented below. Using
SUPRPIC interactively the user selects a particular command by
typing the arguments listed in parenthesis. Note that each argu-
ment is a pair of integers. The first pair of integers designates
the command itself and the other pairs provide the required parameters. As mentioned above the command string 0409010902 is
entered in order to equate the contents of two image arrays. The
integers 04 point to the COPY command, while the integers 0901
point to the first image buffer in logical unit 9 and the numbers
0902 point to the second image buffer in the same logical unit.

              COMPRESS (01,FILE,MODE,TH1,TH2,TH3,TH4)
              MSKHIST (02,FILE,LU,BUF...)
              ERASE (03,LU,BUF)
              COPY (04,LU1,BUF1,LU2,BUF2)
              INVCOPY (05,LU1,BUf1,LU2,BUF2)
              AND (06,LU1,BUF1,LU2,BUF2,LU3,BUF3)
              OR (07,LU1,BUF1,LU2,BUF2,LU3,BUF3)
              EXOR (08,LU1,BUF1,LU2,BUF2,LU3,BUF3)
              AUGRED (09,LU,BUF,CYC,FAC,CNUM,MODE,BORD)
              MRGPLOT (10,LU)
              MIXPLOT (11,LU)
              GRAPLOT (12,LU)
              COLOR (13,LU,BUF,RED,GREEN,BLUE)
              DISPLAY (14,MODE,LU,BUF)
              PICK (15,MODE,LU1,BUFL,LU2,TIME,COLOR)
              STATUS (16)
              ANALYZE (17)
              COUNT (18,MODE,LU,BUF)
              HISTORY (19)
              QUIT (20)

                         SUPRPIC MENU

The command structure of SUPRPIC is as follows:

Utilities:

STATUS - This command presents the user with the status of all program constants and error conditions.

HISTORY - This command presents the user with a list of the last
16 commands executed by the system.

Image Display:

DISPLAY - This command has three modes which permit: (1) display
of a logical image, (2) the display of a graylevel image, (3) the
display of a 3-color image. All images are 512x512 and are displayed at 1-bit per picture element, 8-bits per picture element,
or 9-bits per picture element, respectively.

PICK - This command permits the user to prepare a "ground truth"
map as an overlay to a 512x512 image indicating the locations of
up to 4 classes of objects.

PLOT - This command is used for generating hard copy images in one
of three forms: (1) four 1-bit images (MRGPLOT), (2) four 1-bit
images plus a combination image giving 5-gray-level representation
(MIXPLOT), (3) full-gray-level (16 levels) image using a 4x4 dot-
matrix per picture element (GRAPLOT).

Arithmetic Operators:

ERASE - This command sets all elements of a 512x512 matrix to 0.

COPY - This command equates all elements of one 512x512 matrix to
those of another matrix.

OR/AND/EXOR - These are the standard Boolean operators which are
used to perform image algebra.

Geometric Manipulation:

COLOR - This command expands a 1-bit per picture element matrix
into a 8-bit per picture element matrix using the color intensities specified.

Image Transforms:

AUGRED - This is the major cellular logic command of SUPRPIC and
functions in six modes: (1) cellular logic over an 8-neighborhood
without regard to subfields, (2-5) the same operation selectively
in the first, second, third, or fourth subfields only, (6) cellular
logic over the 8-neighborhood with a subfield order of 1-3-4-2,
(7) the same operation as in (6) with the retention of residues,
i.e., binary 1s whose 8-neighborhood contains all 0s.

Image Measurement:

COMPRESS - This command thresholds a 256-level image at four designated thresholds after histogram normalization. If MSKHIST is
entered prior to this command, then a histogram is constructed only
for those pixels masked by binary 1s.

COUNT - This command counts the total number of binary 1s in a
designated image matrix.

3.   EXAMPLES OF USING SUPRPIC

SUPRPIC has been used primarily in the study of television
images derived from tissue biopsies. Such biopsies are prepared
by sectioning tissue with a microtome to produce a thin slice of
tissue a few micrometers thick which are mounted on microscope
slides. The tissue slice is usually stained with certain biochemical dyes in order to provide information on the chemical composition of tissue constituents, such as cell nuclei, cytoplasmic
granules, connective tissue structure, etc. These dyes selectively
color the tissue and provide a colorimetric labeling. The tissue itself may be drawn from an excised organ or from the percutaneous
extraction of a tissue core with a biopsy needle. The biopsy needle
is hollow and the core which it extracts is approximately 10mm long
and 1mm in diameter.

Examination of the tissue biopsy is frequently performed in
biomedicine in the hospital pathology laboratory. The work of the
pathologist in examining tissue biopsies is characterized not only
by the variety of tissues examined but also by the complexity of
the analysis required. The task of the pathologist is comparable
in difficulty to that performed by the radiologist examining the
x-ray film or x-ray tomogram. In the paragraphs which follow
examples are provided of computer processing of the biopsy image.

3.1 Digitizing the Image

Images were digitized for Carnegie-Mellon University by the
Biomedical Image Processing Laboratory of Jet Propulsion Laboratory
(Pasadena, California) using the Automatic Light Microscope System
model 2 (ALMS2). Each image was digitized in a 512x512 format at
three 8-bit bytes per picture element using the standard color
separation filters employed in the television industry. Picture
element (pixel) spacings of 0.8 micrometers were employed yielding
an image size of approximately 0.5x0.5mm. All images were recorded
on 2400 foot reels of 9-track magnetic tape with each tape holding
ten 3-color images (about 8 megabytes) along with 6 calibration
images (2 at each color) for purposes of monitoring geometric scale,
registration, and color uniformity. In order to achieve the highest
possible quality in the images the ALMS2 was run at a very low
frame time (5 minutes) with a dwell time per pixel of approximately
1ms with the associated photomultiplier 3db bandwidth set at 500Hz.
This mode of operation permitted the recording of a full magnetic
tape in approximately half a day. Standard ALMS2 headers are

provided with each image and images were separated with end-of-file
marks. A typical image is shown in Figure 2 (and in Color Plate
23).* This image is of human liver exhibiting fatty change throughout as well as infectious infiltrate in the lower left quadrant as
well as incidental mild fibrosis in the lower right quadrant.

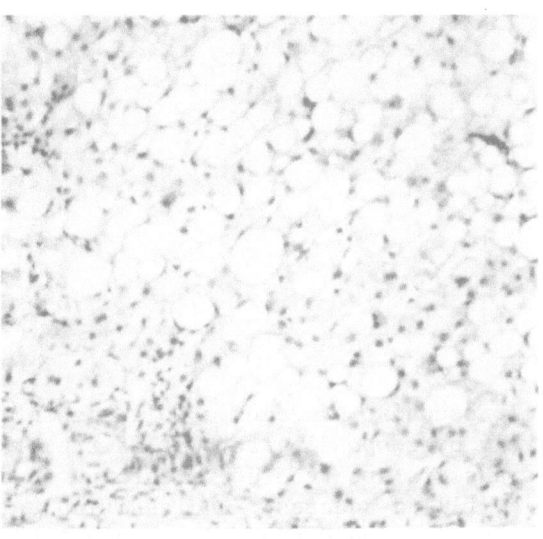

Fig. 2. This 512x512 grayscale image of fatty human liver has
dimensions 0.5x0.5mm. It shows fatty change throughout, infectious
infiltrate in the lower left quadrant, and incidental mild fibrosis
in the lower right quadrant. (See also Figure 1.)

The first step in performing an interactive computer analysis
of an image such as that shown in Figure 2 is to determine the
identity of the various cells which constitute the tissue structure.
In order to do this interactively the computer is instructed with a
liberal cell-nucleus finding algorithm. In this particular case a
13-step sequence was used having 12 identical CLOs whose arguments
were given by (CYC=6,FAC=5,CNUM=4,MODE=6,BORD=0). This CLO was
applied to each of the logical images shown in Figure 1. The final
step of the algorithm was to logically OR all 12 results together.
In the case cited the result was a logical array containing approximately 900 residues marking the location of candidate nuclei. All
of these candidate nuclei were then re-examined using the microscope with the result being the reference map presented in Figure 3.
Nuclei were categorized into those belonging to cells which could
be identified with a relatively high degree of certainty and those

---

*The color plates will be found following page 144.

A LIVER TISSUE "GROUND TRUTH" MAP -SUPRPIC- 14 NOV 79

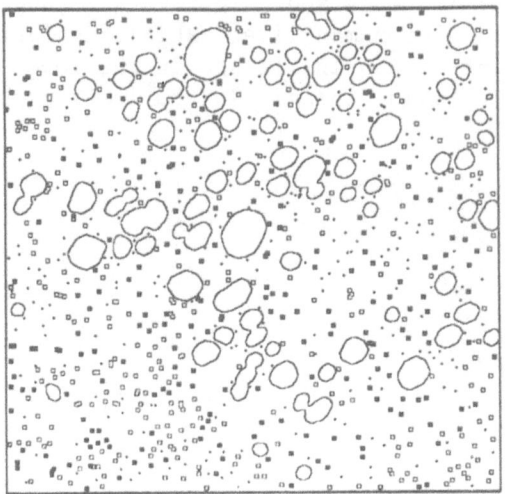

Fig. 3. The microscopist produces a cell classification reference
map interactively by responding to queries from the computer. The
microscopist's answers are recorded according to whether the cell
is classifiable (filled squares); unclassifiable (open squares);
not a cell (crosses). Boundaries of fatty cells are shown as well.

nuclei belonging to cells whose type was difficult to determine.
Most of the cells which could be typed (approximately 250) were
hepatocytes and leukocytes. Because of the relatively large number
of these cells and the small number of other types (approximately
30), the study was continued with a goal of describing the "architecture" of the location of these cell nuclei. A color-coded
reference map giving further information is shown in Color Plate
24.

3.2 Determination of Tissue Image Architecture

     The computation which has been found to best elucidate tissue
architecture is the generation of the graph called the exoskeleton.
The concept of an exoskeleton was originally introduced by Prewitt.
Prewitt also introduced the concept of the endoskeleton which has
also been investigated by Hilditch and Montanari. One method of
generating the exoskeleton is to use the SUPRPIC AUGRED command
with the parameters (CYC=N,FAC=5,CNUM=4,MODE=6,BORD=1) acting on
the 1s complement of the array of residues. The result of this
command is to cause the expansion of each residue in the image
plane until its expansion is arrested by collision with other

expanding residues. Color Plates 25 and 26 provide examples of
the growth of this graph. In Color Plate 25 growth is arrested at
various steps in the process and color coded at this point. As can
be seen, some regions enclosed by the exoskeleton grow to their
full extent early in the computation because their growth is immediately arrested by their close proximity to other residues. As
these regions grow the boundaries between regions (which are the
arcs of the exoskeleton) are completed. Color plate 26 is another
representation of this fact. In this Color Plate the arcs themselves are color coded according to whether they are completed
early in the calculation (white), later in the calculation (blue),
or at the end of the calculation (green). It has already been
noted by Preston et al. (1979) that the arcs which are color-coded
white point to regions of high cell density and, if they are non-
uniformly distributed within the tissue image, indicate infectious
infiltrate. This is further illustrated in Color Plates 27 and 28
which were derived from images of normal and abnormal (infected)
kidney tissue, respectively. In this case the image size is 1x1mm.

A composite showing both the endoskeleton and the exoskeleton
is shown in Figure 4 and an example of the location of short arcs
in the endoskeleton in Figure 5. As can be seen the endoskeleton
is less useful than the exoskeleton in locating regions of infection although there is a definite location of short endoskeleton
arcs in the region of infectious infiltrate.

A complete tissue architecture program has been coded in the
SUPRPIC command language for use in both locating cell nuclei, fat
droplets, vessels, and in calculating the exoskeleton. Using the
COLOR command the results of this program produce a color cartoon
of the tissue. This color cartoon shows the cell nuclei as laven-
der, fat droplets and vessels as white, and the exoskeleton in
orange. This command sequence was applied to further examples of
both normal and abnormal liver tissue with the results shown in
Color Plates 29 through 32. These color cartoons are currently
being analyzed numerically by computing the probability density
function of the cell nuclei density, the size of the polygons
described by the exoskeleton, and the variance and kurtosis of
these quantities. The intention is to relate this numerical data
to the disease state. Two Masters projects have been pursued along
these lines by students Castillo (1978) and Yorkgitis (1979) who,
for a limited set of images, were able to show that the computer
could differentiate between normal and disease states and, furthermore, differentiate between those specific disease states characterized by acute viral hepatitis and alcoholic hepatitis for a
limited number of patients. Although results are too early to completely evaluate, this approach appears to show great promise of
success.

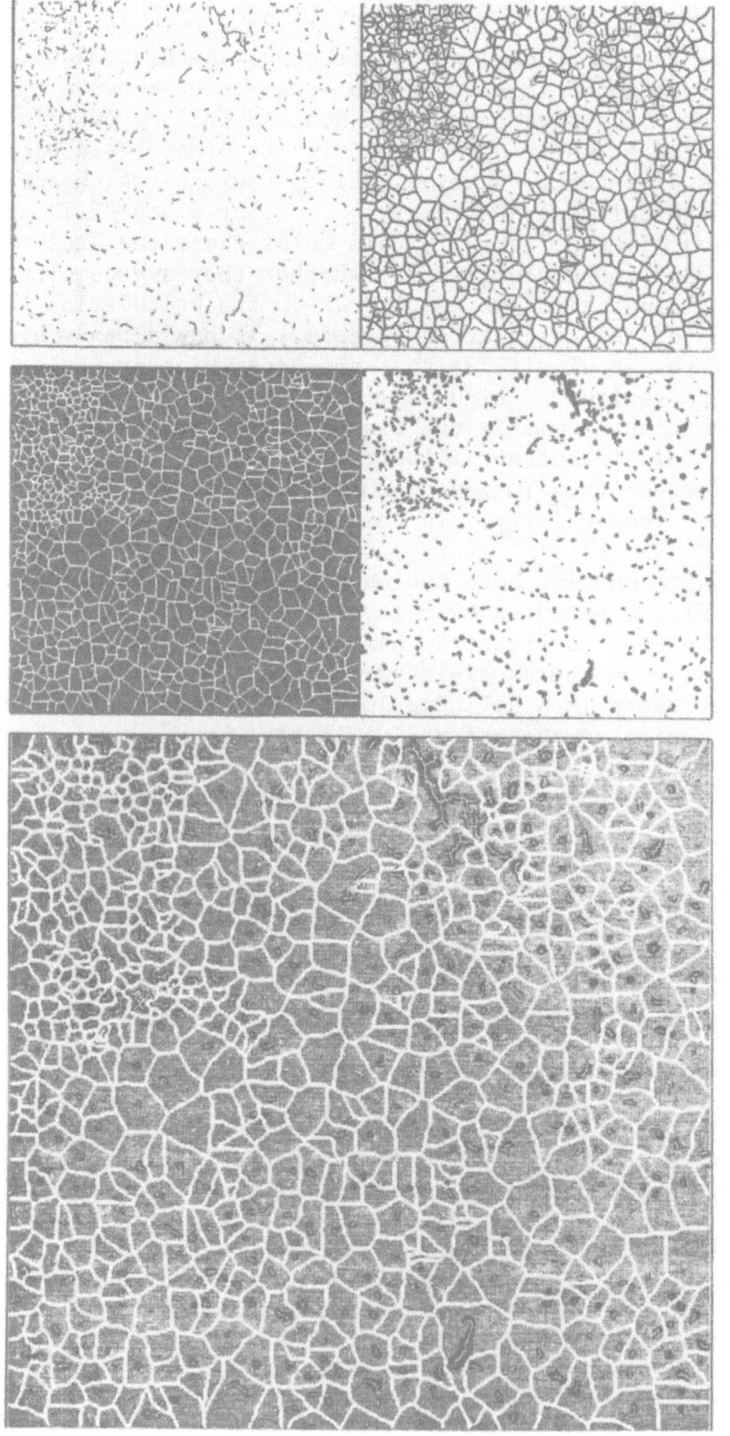

Fig. 4. The SUPRPIC interactive image processing system generates a composite of four logical images when the MIXPLOT command is executed. Here the images merged are: (1) complemented exoskeleton, (2) endoskeleton, (3) threshold 08 image, (4) ORed endoskeleton and exoskeleton. A complete analysis of the tissue image architecture is the result.

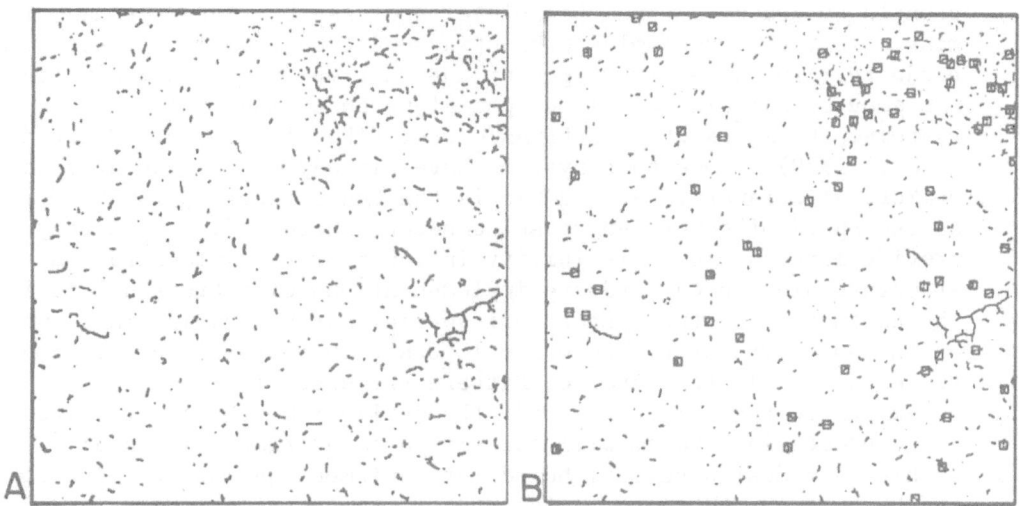

Fig. 5. By locating and marking (open squares in B) short arcs of
the endoskeleton (A), the degree of local infection may be measured.

4.   SUMMARY AND CONCLUSIONS

The analysis of medical images is an inexact science. Unlike
such systems as communications networks, mapping radars, space
vehicle control and guidance, etc. the mathematical basis for image
analysis is undeveloped. We know that it is important to segment a
scene by locating boundaries and, within the segments found, to
locate, detect and classify specific objects. The mathematical
equations to be solved, however, cannot be written. The user must
visually observe the two-dimensional function or ("image") with
which he is dealing not only in its initial state but also as it
is being processed. Because images are characterized by large
number arrays (often 1 megabyte or more), the computational burden
on the computer when required to provide instantaneous results is
extreme. Because of this requirement for high-speed interaction,
image processing systems are quite different from the ordinary
general purpose computer installation. It is important in struc-
turing these systems to use every facility of the modern computer
to gain speed of computation.

The designer must be fully aware of limitations of disk input/
output, core direct memory access rates, and data transmission over

the various communication buses of his system. Software must frequently be coded in mixed high-level and assembly language and, in many cases, time-consuming subroutines must be translated into hardware. In the early stages of interactive image processing this was done by building complete special-purpose image analysis systems. Examples were CELLSCAN/GLOPR, PIP, PICAP, etc. (Preston et al., 1979). Some such systems have already been translated into commercial hardware. An example is the Coulter/Perkin-Elmer diff3. As the speed of general-purpose computers increased, and as they provide more and more flexibility in incorporating standard bus structure high-speed hardware implementations of subroutines it is the estimation of the author that special purpose peripherals such as the Golay Transform Processor (GLOPR), the Pattern Information Processor (PIP), the Parallel Pattern Processor (PPP), etc. will no longer be required. Instead, the designer of the interactive image processing system will purchase a main frame commercial machine and design several boards of equipment (along the line of the Perkin-Elmer ULI) to implement certain specific subroutines within his image processing package. Such a development is already taking place and it is believed will be characteristic of the 1980s.

5.   REFERENCES

Castillo, A., "Automatic Computer Identification of Liver Biopsies," Master's Dissertation, Depart. Elec. Engrg., Carnegie-Mellon Univ. (1978).

Golay, M. J. E., "Hexagonal Parallel Pattern Transformations," IEEE Trans. Comput. C-20:551 (1971).

Preston, K., Jr., "Use of the Golay Logic Processor in Pattern Recognition Studies Using Hexagonal Neighborhood Logic," in Computers and Automata, New York, Polytechnic Press (1972), pp. 609-624.

Preston, K., Jr., Duff, M. J. B., Levialdi, S., Norgren, P. E., and Toriwaki, J-i., "Basics of Cellular Logic with Some Applications in Medical Image Processing," Proc. IEEE 67(2): 826-856 (1979).

Preston, K., Jr., and Norgren, P. E., "Interactive Image Processing Using the Golay Logic Processor (GLOPR)," Nikkei Electronics (26 Feb. 1973), pp. 106-123 (in Japanese).

Yorkgitis, D. P., "Automatic Computer Classification of Human Liver Biopsy Digitizations from Two Different Diseases and the Normal State," Master's Dissertation, Dept. Elec. Engr., Carnegie-Mellon Univ. (1979).

REAL-TIME IMAGE PROCESSING IN AUTOMATED CYTOLOGY

R. Suzuki and S. Yamamoto

Central Research Laboratory

Hitachi, Ltd., JAPAN

1. INTRODUCTION

Uterine cancer is detected in its early stages by screening cells scraped from the uterus and spread on a microscope slide for visual observation. Cells are roughly screened by a cytological technician called a "screener." A physician then makes the final diagnosis. Cyto-screening has been automated by us using image processing and pattern recognition techniques. The authors have made significant improvements in the following factors: (1) screening accuracy, (2) screening speed, and (3) sample preparation. New cytologic feature parameters and a hierarchical classification algorithm have been developed. As feature parameters we use nuclear area, nuclear area/cytoplasm area, and nuclear area with higher density. The hierarchical classification algorithm incorporates the cell growth method of Suzuki (1978). Screening speed is expected to be twice that achieved in manual screening. The authors aimed at a speed of 5 min./sample and have succeeded in developing a system with a speed of 2 min./sample.

2. SYSTEM DESIGN AND SPECIFICATIONS

This section describes the functions of the major system components and provides system specifications.

2.1 System Design

The newly developed cyto-screening system is shown in Figure 1 and its block diagram is provided in Figure 2. Overall control

Fig. 1.   The Hitachi cyto-screening system.

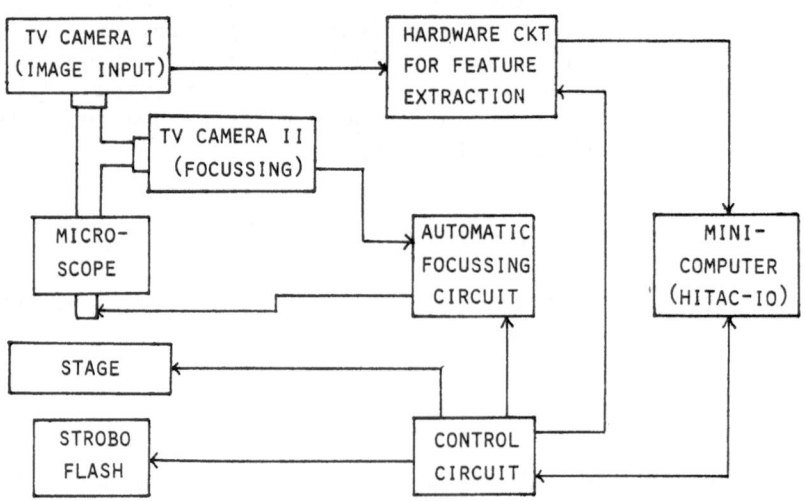

Fig. 2.   Block diagram of the system.

plus the image processing and recognition functions are provided
by a mini-computer (HITAC-10). Images of the specimen are generated
by light from a strobe flash using the microscope to magnify the
specimen (10x objective lens plus a 5x ocular). The microscope
stage holds and moves the specimen. Input images are formed by TV
camera I and feature extraction is performed in hardware. The
circuit used for feature extraction detects cells and extracts
feature parameters in real time. Input images are automatically
focused using TV camera II and an automatic focusing circuit.

2.2 System Specifications

Specifications of the system are shown in Table 1.

Table 1 - Specifications of Automatic Screening System

Items	Specification
Screening Speed	2min/1 sample (20mm x 20mm by 1.3um pitch)
Image sensor	TV camera
Light	Strobo flash lamp (Xe)
Feature extraction	2-D parallel processor
Automatic focussing	Continuous focussing by three images (using two TV cameras)
Feature parameters	NA, N/C, NAH
Classification algorithm	Hierarchical classification incorpora- ting cell differentiation

A screening speed of 2 min./sample has been realized by using a TV
camera as an image sensor. The TV camera has a 300x400um field of
view with 1.3um line spacing and a processing speed of 2 fields
each 1/30 second. To achieve this screening speed, three new
methods are employed: (1) video image generation using a strobo
flash, (2) continuous focusing, and (3) real time feature parameter
extraction.

Due to image generation by strobo flash, image input of every other field has been realized. Continuous focusing has made the waiting time short to the moment when the focused image is ready. In conventional systems, an image has to be input more than once. The newly developed system can process three kinds of images at once and, as a result, focusing and image input can be carried out simultaneously. Feature parameters are extracted in real time using an image memory composed of shift registers. Every cell in the image is processed continuously.

3.   IMAGE INPUT

This section describes the design and use of the system which generates input images for the automated cyto-screening system.

3.1 Image Input System Design

The image input system is block diagrammed in Figure 3. It is composed of five blocks whose components are: (1) a light source to illuminate the microscope field, (2) a moveable stage to feed and transport specimens, (3) the microscope to magnify portions of the specimen, (4) the image sensor which generates the video signals, and (5) the automatic focusing module which maintains continuous focus. In the discussion below the blocks are grouped as follows: (1) image sensor module (image input and light source), (2) screening module (moveable stage and microscope), and (3) automatic focusing module.

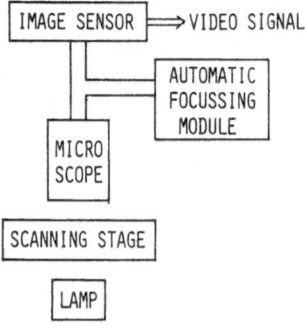

Fig. 3.   Block diagram of image-input module.

3.2 Image Sensor Module

A TV camera has been selected as an image sensor in comparison with the flying spot scanner and the solid state sensor by taking into consideration cost, sensitivty, and wavelength characteristics. As a TV camera tube, the Charnicon E5072 is used due to its high sensitivity in the red. This tube is not ideal for after-image elimination and random scanning, but these defects can be solved by an appropriate image sensing method and stage moving method.

Ordinarily high speed image input using TV cameras produces a degeneration in image quality due to after-image effects and stage motion. It requires 5 to 6 frame times before an image can be input since 3 to 4 frame times are required for the stage to settle. Two more frames are required for the after-image to fade away. In order to solve this problem a strobo flash illumination method has been adopted for image input. This method pulses the strobo flash exactly at the start of the video signal and inputs an image in the following frame. The after-image is eliminated in the next few frames. The method has been implemented efficiently using the Charnicon E5072 and the Sugawara MS 210 strobo flash. This is a 5 watt xenon strobo flash of the light concentration type.

Images taken by strobo flash are shown in Figure 4. This shows that by using an appropriate amount of light an effective image is produced and the after-image is negligible. In fact the after-image is reduced to 10% of the primary condition in one frame time. Accordingly, we input images every other frame time.

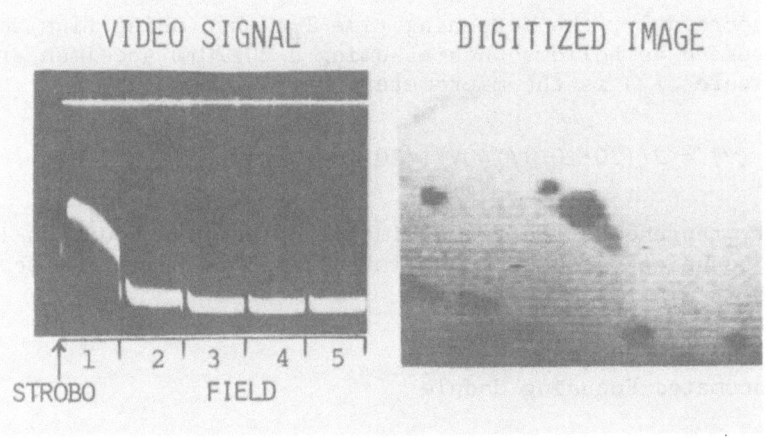

Fig. 4. Example of an image produced by strobo-flash.

3.3 Screening Module

The movement of the specimen, namely, the method of moving the
stage is called the "screening method." Screening is divided into
two methods: (1) fine screening and (2) the combination of fine
and rough screening. Fine screening is the method used when every
part of the specimen is screened with the same fine scanning pitch
and the other method is to screen with a rough scanning pitch and
then with a fine scanning pitch those parts of the specimen which
are suspicious.

Both methods have been evaluated from the point of screening
speed and the complexity of the optical, electrical and mechanical
systems. The combined method has an advantage in screening speed
while fine screening has an advantage in requiring simpler optical,
electrical and mechanical systems. A prototype of the mechanical
stage for the combined method has been produced for trial but it
has been difficult to get the system to meet specifications.
Therefore fine screening has been adopted, regardless of its long
scanning time, for its simple mechanical system and constant image
input ability. Fine screening has the additional advantage that
the mechanical system can drive the stage to one direction continuously.

Suppose the field of view contains vertically 240 and hori-
zontally 320 effective picture elements and the scanning pitch with
respect to the surface of the specimen is γ, the effective window
W_γ with respect to the specimen is expressed by equation (1)

$$W_\gamma = 240\gamma \times 320\gamma . \tag{1}$$

Accordingly, the screening time T_p, when using fine screening,
is expressed as follows when assuming a 20x20mm specimen area and
the measure of γ is the micrometer.

$$T_p(\gamma) = 2((20\cdot10^3)/240\gamma)((20\cdot10^3/320\gamma)t_f \tag{2}$$

where t_f represents the frame time (1/60 sec.). Equation (2) shows
that a scanning pitch of 1.3um is required to realize a scanning
speed of 2 min./specimen.

3.4 Automated Focusing Module

The focal depth of a microscope depends on the numerical
aperture of the objective lens. Automated focusing is necessary
for obtaining satisfactory images in cyto-screening and is usually
carried out by comparing images obtained from both up and down

positions of the stage. Therefore, to make an image which is in best focus, at least three images have to be input. This requires mechanical motion which decreases the processing rate. A method is needed to detect best focus without mechanical movement. Two methods can be used: (1) the distance measuring method and (2) the image processing method.

In the distance measuring method, distance between the sample surface and the objective lens is kept constant. A sample used in cyto-screening is prepared by placing a layer of cells between the microscope slide and a cover glass. If the distance between the surface of the cover glass and the slide changes by more than 10um, it proves to be impossible to focus even if the distance between the surface of the cover glass and the objective lens is kept constant. Therefore, we must use the other method.

When the distance between the sample surface and the objective lens of magnification m is changed minutely by $\Delta \ell$, the movement ΔL of the image is expressed as follows:

$$\Delta L = \Delta \ell \times m^2 . \tag{3}$$

This suggests that the best focused image can be obtained by changing the ocular position by ΔL. During this process, magnification and aberration will change a little, but the effect will be negligible in practical use if the $\Delta \ell$ is of the order of the focal depth.

When the actual optical path between objective and ocular is L, oculars of the same magnification placed at the positions (L-ΔL, L, L+ΔL) will form images equivalent to those obtained with the objective lens at the positions (ℓ-$\Delta \ell$, ℓ, ℓ+$\Delta \ell$). Using this method, images will be obtained at the same time and can be compared without mechanical change to detect the focal condition.

Realization of this method with two TV cameras is shown in Figure 5. An image generated with optical path length L is input through TV camera I, those with optical path lengths (L-ΔL) and (L+ΔL) through TV camera II, and the images are compared electronically to detect the focal condition. Optical path length differences between the images sensed by TV camera II are easily realized by using an optical flat. To fully realize this optical method would require exact optical registration techniques. This is difficult for practical use. The method adopted in our system is to detect the focal condition using macroscopic information on the whole image which works even though the images to be compared are several picture elements out of registration.

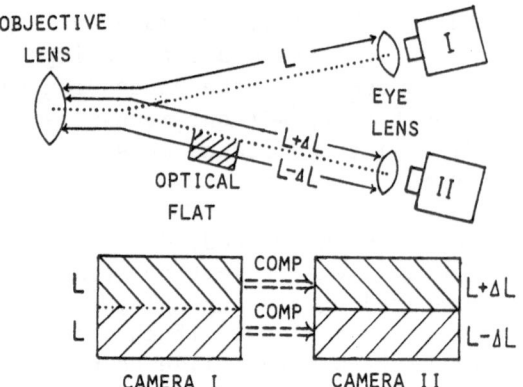

Fig. 5. Optical schematic of auto-focus technique.

4.   FEATURE EXTRACTION METHOD

    We have adopted a hierarchical recognition algorithm for the
cell recognition process. In this method, NA (nuclear area), N/C
(nuclear area/cytoplasm area) and NAH (nuclear area with higher
density) are used as feature parameters. The cell division
process is recognized by NA and N/C, and re-recognition is carried
out using NA, NAH according to the cell growth process. To opti-
mize circuit performance NA, CA (cell area), and NAH are calculated
by feature extracting hardware and N/C by the minicomputer. Since
image input by strobo flash is used detection of every cell in the
field and detection of their features must be carried in only one
frame time.

    The optical density profile of one line through the center of
a cell is shown in Figure 6. The necessary density thresholds
which must be established in going from cell detection to feature
parameter extraction are as follows: (1) G_{min}: minimum background
density, (2) TH_c: threshold density for cell detection, (3) TH_{ca}:
threshold density for CA (cell area), (4) G_{max}: maximum density
in nucleus, (5) TH_{na}: threshold density for NA (nuclear area),
(6) TH_{nah}: threshold density for NAH (nuclear area with high
density).

4.1 Feature Extraction Hardware

    Feature extraction requires 2-dimensional parallel processing.
Taking into consideration the scale of the system and the processing
speed, the method adopted is to process the whole field by applying
local 2-dimensional parallel processing. The structure of the
hardware is shown in Figure 7. A 6-bit, 6.14 MHz, A/D convertor
is used to feed a 2-dimensional shift register memory:

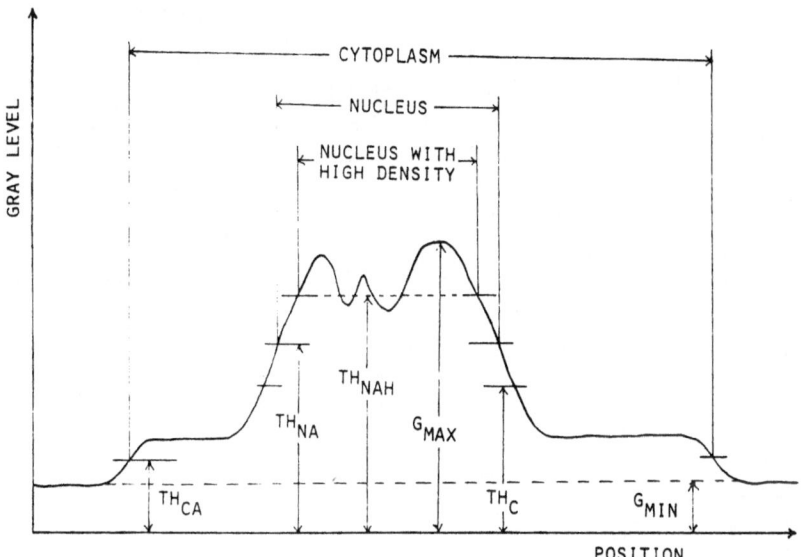

Fig. 6. Thresholds determined from one video scan across a cell (TH: threshold for cell area; TH_c: detection threshold; TH_{na}: threshold for nuclear area; TH_{nah}: threshold for high density nuclear area).

6^{bit} x $320^{P.e.}$ x 21^{line} where p.e. stands for picture elements. A thresholding circuit outputs 21 lines of binary signal at the threshold levels TH_c, TH_{ca}, TH_{na}, and TH_{nah}. The cell extraction circuit finds cells and detects the center of each cell by means of the binary signal generated at threshold TH_c. The area calculation circuit computes the area of cells using binary signals generated at thresholds TH_{ca}, TH_{na}, and TH_{nah}. The minimum value detecting circuit finds G_{min} for every line. A second 2-dimensional memory (6^{bit} x $8^{P.e.}$ x 3^{line}) is used to calculate TH_{na}. At the same time the maximum value detecting circuit finds G_{max} and the cell threshold calculating circuit computes the threshold within every area as follows:

$$TH_c = G_{min} + \alpha \qquad\qquad (4)$$

$$TH_{ca} = G_{min} + \beta \qquad\qquad (5)$$

$$TH_{nah} = (TH_{na} + G_{max})/2 \qquad\qquad (6)$$

where α and β are constants and TH_{na} is found from the differential of 3 lines at the center of the nucleus. Finally the interface transmits data to the mini-computer and sends out and receives control signals to the various control circuits.

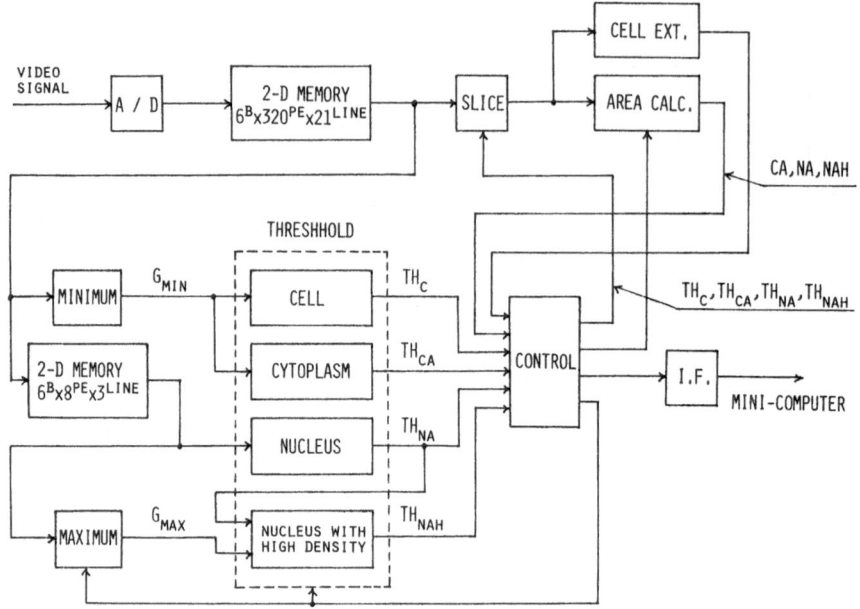

Fig. 7.   Block diagram of the feature extraction module.

4.2   Feature Extraction Procedure

The digitized 6-bit image signal is transmitted into the major 2-dimensional memory. The binary pattern for each cell extracted is generated by calculating G_{min} using the output of this memory and by transmitting TH_C to the thresholding circuit. Existence of a cell is determined by sending this binary pattern to the feature extraction circuit. Feature extraction is then carried out after memorizing the position of the detected cell.

The feature parameter extraction sequence is shown in Figure 8. TH_{ca} is calculated at the same time a cell is detected (during the 1st HD interval). During the 2nd HD interval, CA is calculated by the area calculation circuit using as input the binary pattern generated by TH_{ca}. At the same time, G_{max}, TH_{na}, and TH_{nah} are calculated by threshold detection circuits. During the 3rd HD interval, NA is calculated using TH_{na}. During the 4th HD interval, NAH is calculated using TH_{nah}. Finally, these feature parameters are transmitted to the mini-computer.

Adopting this method, the feature parameters of a cell are extracted during three HD intervals (about 250us). At this time all the cells in a field are ready to be processed. Furthermore, the structure of the circuit is simplified by the fact that the same kind of processing is carried out during each HD interval.

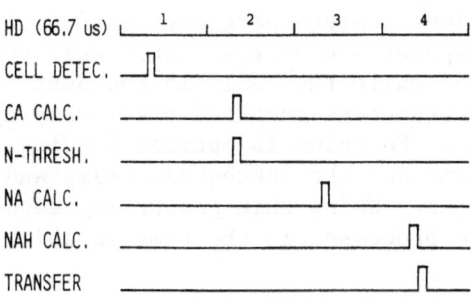

Fig. 8. Timing sequence for parameter extraction.

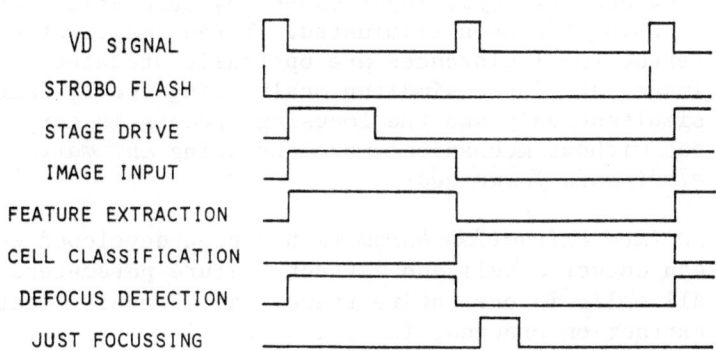

Fig. 9. Timing sequence for the automatic cyto-screening system.

5.   PROCESSING SEQUENCE

     The processing sequence of the system is shown in Figure 9.
First, an image is input using the strobo flash just before the
start of VD (Vertical Drive) of the TV camera. The stage is moved
to the next position. The image signal from one frame is input to
the feature extraction hardware. The image is analyzed in real
time. Cell detection, feature parameter extraction, and data transfer to the mini-computer are carried out. Cell recognition is done
by the mini-computer while the cells in the next frame are analyzed
using the feature parameters supplied previously by the feature
extraction hardware. Focusing is optimized using the image signals
obtained in one frame and the subsequent frame and focusing adjust-
ments are carried out. Using this processing sequence, all cells
in one frame can be processed in the time in which two images are
input (1/30 sec.).

6.   CONCLUSION

     A high-speed, cyto-screening system has been developed. The
system has a processing speed of 2 min./specimen. The following
three systems have been adopted to realize this speed.

     (1) *Image sensor system* using strobo flash to reduce any
          slow down in image input speed due to stage vibration
          and the after-image produced by the TV camera tube.
          Images are input every 2 frame times (1/30 sec.).

     (2) Slow down of image input caused by automatic mechanical
          focusing has been eliminated. Three images at dif-
          ferent focal distances are optically produced. These
          images are input simultaneously using two TV cameras
          simultaneously and the focusing process is carried
          out without mechanical movement using *automatic
          electronic focus system*.

     (3) *Feature extraction hardware* has been developed which
          can detect a cell and extract feature parameters for
          all cells in one entire frame. As a result, feature
          extraction proceeds in real time.

     These newly developed techniques have made it possible to
screen each microscopic image in the time used to input two images
(1/30 sec.) and to screen a 20x20mm specimen in two minutes.

     This automated cyto-screening system will be put to field test
to examine screening accuracy in the future. The remaining problem
is to simplify sample preparation to match this system.

7. REFERENCES

Suzuki, R., "Automated Classification of Uterine Cancer Cells
    Utilizing the Concept of Cellular Differentiation," Proc.
    4th Joint Conf. on Pattern Recog. (Kyoto, 1978).

THE DEVELOPMENT OF A NEW MODEL CYTO-PRESCREENER FOR CERVICAL CANCER

S. Watanabe, S. Tsunekawa, Y. Okamoto*, I. Sasao*,
and T. Tomaru**

Toshiba Research and Development Center

Saiwai-ku, Kawasaki, 210 JAPAN

1.   INTRODUCTION

Mass cancer screening of cervical smears based on manual classification of microscopic specimens is a very expensive procedure. Therefore, large research efforts have been made to realize cytology automation. We developed a cyto-prescreener called "CYBEST" (Cyto-Biological Electronic Screener by Toshiba) in 1972 and have conducted various kinds of feasibility tests as described in Watanabe (1976). Based on experimental results obtained through these tests, we have developed a new high-performance cyto-prescreener. In the following section the fundamental ideas which have been developed to improve the performance of a practical system are described.

2.   FUNDAMENTAL IDEAS

The following four technical ideas were taken into consideration in developing the new cyto-prescreener. These ideas have contributed to realizing a practical machine having better cost-effectiveness.

---

*Toshiba Nasu Works
**The Medical Information System Development Center

2.1 Improvement of Diagnostic Performance

In the new cyto-prescreener the diagnostic performance has been improved to yield less than 30% false positives and less than 3% false negatives. Because of technical problems in the old model, its best performance rates were limited to 40% false positives and 5% false negatives. In order to meet the requirements set by the medical side, the new cyto-prescreener has employed several design concepts which are described below.

2.2 Improvement of Processing Speed

The new cyto-prescreener can process 200 microscope slides automatically in 10 hours whereas the old model took 6 minutes to process one slide. This still does not satisfy the current demand for mass cervical smear screening.

Specially designed electronic circuits and related mechanics permit high-speed processing at 3 minutes per smear. This permits an all-night operation to be run at a rate of 200 smears per night as well as an additional 200 smears per day.

2.3 Automatic Slide Preparation

We have also developed an automatic apparatus to prepare the slides which produces a uniform smear in terms of the number of cells per smear, cell dispersion, staining condition, etc. Highly stable characteristics of the smear are essential requisites for obtaining better diagnostic performance.

Table 1 compares the features of the new cyto-prescreener with the old model. The new model is superior to the old one especially in its screening speed, screening performance, and slide magazine capacity.

3. IMPROVEMENT OF DIAGNOSTIC PERFORMANCE

We have concentrated our efforts on improving the diagnostic performance by making the following technical improvements.

3.1 Improvement of Cell Image Quality

We have effected several improvements in the image data input section in order to input a high quality cell image with 1μ resolution. The input section now employs a high resolution TV camera

Table.I   Comparison with CYBEST/72 and New Model

System / Item	CYBEST/72	New Cyto-prescreener
Input device	FSS	TV Camera and stroboscope
Spatial resolution	4μ (low) 1μ (high)	2μ (low) 1μ (high)
Scanning speed	80msec/field	80msec/field
Image processor	Mini-computer	Special purpose computer
Screening rate	About 60%	Greater than 70%
Screening speed	6 minutes/slide	3 minutes/slide
Slide magazine capasity	50 Slides	200 Slides
Sample preperation	Manual work	Semi-automatic work

and its associated optical system. These improvements allow the new cyto-prescreener to resolve the chromatin structure within the cell nucleus which makes cell diagnosis more effective.

3.2   Addition of Morphological Feature Parameters

A new shape parameter C_S has been added as a morphological feature parameter in addition to cytoplasmic area C_A, nuclear area N_A, nuclear optical density N_D, and nuclear shape N_S. Another new feature called "the number of blocks in the nucleus" N_B is obtained from the nucleus image thresholded at a suitable level. This new feature parameter represents the nucleus presence as well as the chromatin structure. The shape parameter C_S is expressed as the square of the boundary length of the cell divided by the cellular area and is used mainly for discriminating a cervical cell from the leukocytes, mucus, and other artifacts. Examples of the new feature parameters of typical normal and abnormal cells are shown in Figure 1.

3.3   Improvement of Preprocessing Program for Feature Extraction

The values of the features extracted depend on the segmentation of the cytoplasmic and nuclear regions. The feature parameters can be easily determined by simple logical operations if the segmentation is correctly performed on a given cell image.

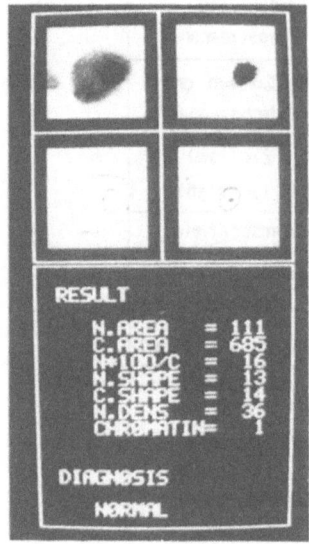 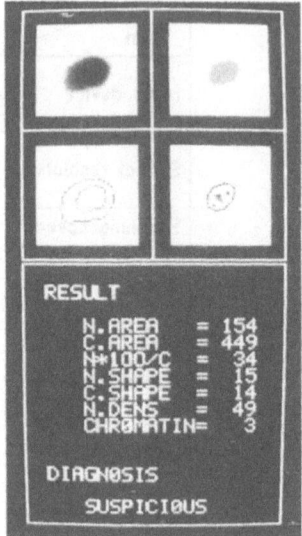

Fig. 1. Examples of extracted feature parameters: normal cell
(left); abnormal cell (right).

In the new cyto-prescreener we have employed an efficient segmentation algorithm which analyzes the density histogram of a given
cell image following a top-down approach. Several methods are used
for the segmentation of cells: the differential histogram method
described by Watanabe et al. (1974), the two-dimensional histogram
method of Holmquist et al. (1978), etc. The differential histogram
method is superior since it can distinctly emphasize the region
boundaries but it is not resistant to noise and requires a long
processing time.

The two-dimensional histogram method is especially effective
in analyzing cells found in cervical smears. This method requires
hardware for calculating the probability density information for
the green and red images. Our preprocessing program contributes to
speeding up the execution time and simplifying the hardware implementation. The probability density function is calculated for a
given cell image and then a smoothing operation is performed to
eliminate noise. Local maxima and minima are detected and the number of peaks (local maxima) is computed. The number and location
of the peaks provide data for segmenting the background, cytoplasm
and nucleus regions. Experimental results indicate that the number
of peaks is not greater than five and their relationships to region
boundaries can be classified into eleven types. Table 2 lists
these eleven types and Figure 2 shows its typical example.

Table.2 Type of density histogram of cell

No. of peaks	Type	Relation between peak and region
1	1	B include C and N
2	2	B include C ; N
	3	B ; C include N
3	4	B ; C ; N
	5	B include C ; N_1 , N_2
	6	B ; C_1 , C_2 include N
4	7	B ; C_1 , C_2 ; N
	8	B ; C ; N_1 , N_2
5	9	B ; C_1 , C_2 , C_3 ; N
	10	B ; C_1 , C_2 ; N_1 , N_2
	11	B ; C_1 ; N_1 , N_2 , N_3

(B : Background, C : Cytoplasm, N : Nucleus)

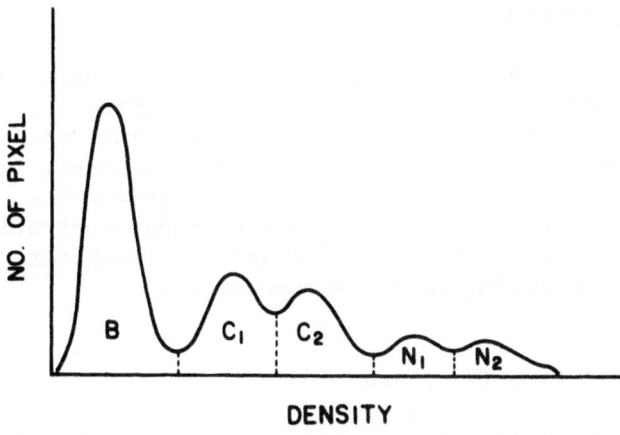

Fig. 2. Example of a probability density function of Type 10.

After the smoothed probability density function is classified, then segmentation is made on the basis of type. Next, judgement is made as to whether an input pattern is compatible with the segmentation result obtained by the selection of its type. If an input pattern is incompatible with the result obtained, another type with the same number of peaks is selected and segmentation is repeated. This process is repeated until an input cell image is correctly segmented into the background, cytoplasm, and nucleus regions. In the course of this preprocessing procedure, the degree of region separability is determined and is used to determine the final diagnostic result for a given cell.

3.4   Increase in Number of Cells Analyzed

By specially designed hardware and efficient programs we have increased the number of cells processed for detailed diagnosis as much as possible. In the old cyto-prescreener, the number was limited to 300 cells per smear due to the considerably long processing time taken by a minicomputer. Multi-stage processing was conducted consisting of first making a coarse diagnosis with 4µ low-resolution to select 300 suspicious cells and, second, to perform a detailed diagnosis on these 300 cells with 1µ high-resolution. Low-resolution selection of suspicious cells, however, did not provide adequate results thus affecting the second fine-diagnosis step adversely. From the statistical point of view it is better to use high resolution throughout to increase the number of suspicious cells detected in the first step (see Moore, 1974).

4.   HARDWARE IMPLEMENTATION

Figures 3 and 4 show the schematic block diagram of the new cyto-prescreener and a photograph of the equipment, respectively. The new cyto-prescreener consists of three major units: (1) the slide transport mechanism and TV microscope scanner, (2) the preprocessing unit and (3) the diagnostic computer unit. Each unit operates under program control; the data input controller, micro-programmable image processor, and diagnostic computer. This new model has the following hardware components.

4.1   High Resolution TV Camera

The new model has implemented a high resolution input device by a combination of the TV camera and strobe illumination in place of the conventional flying-spot scanner included in the old model. This implementation improves not only the S/N ratio but also speeds up cell image input. The vignetting problem and video signal fluctuation incidental to an input scanner of this type are eliminated by the use of a digital shading correction circuit.

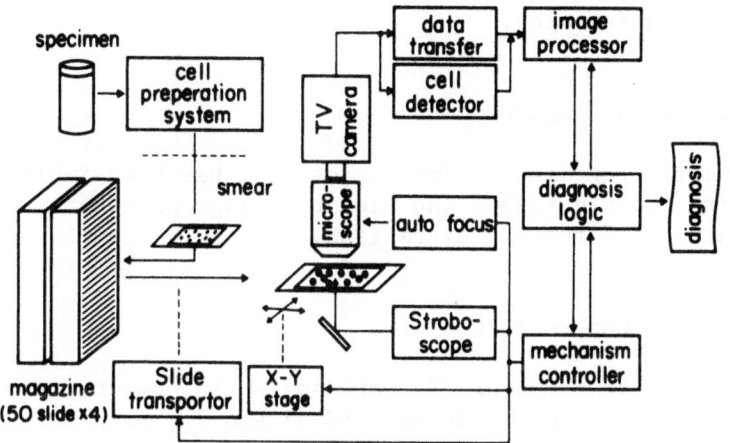

Fig. 3.   Schematic diagram of the new cyto-prescreener.

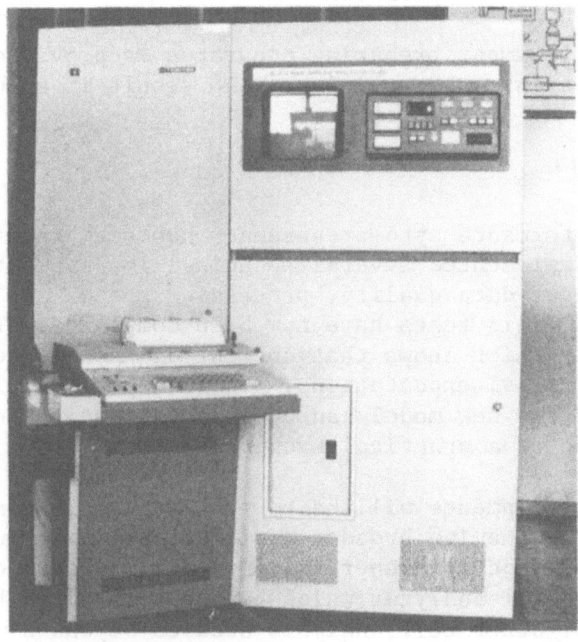

Fig. 4.   Photograph showing the new cyto-prescreener.

4.2 Automatic Focusing Mechanism

A digital circuit for detecting the contrast of a given cell image is implemented to perform automatic focusing. This mechanism improves both image quality as well as permitting the use of new morphological features.

4.3 High Speed Image Processor

A micro-programmable image processor has been implemented to accomplish the high-speed and flexible image processing tasks in segmentation and feature extraction.

4.4 Flexible Controllers

Both a microcomputer (I-8080) and a minicomputer (TOSBAC-40C) are used for data input control and diagnosis, respectively. All information processing is conducted in parallel by three computers: the computer for controlling the data input, the computer for diagnosis and the computer for image processing. By distributing control and computing functions processing speed is improved.

4.5 Automatic Smear Preparation

An automatic smear preparing apparatus is provided for the cell dispersion, smearing, and staining. The result is a uniform smear.

5. CONCLUSION

A high-performance cyto-prescreener has been realized. The new model has implemented several technical improvements in smear preparation, input data quality, processing speed, and diagnostic logic. Several field tests have now been completed. Table 3 is a confusion matrix which shows that current diagnostic performance satisfies our initial expectations. These experimental results have shown that the new model can be used in institutions for mass cancer screening as a practical machine.

Further developments will be to realize a more compact and economical machine having broader applications and to utilize more of the technical function generated in analyzing each cell image in the smear. The cell analyzer which has been realized has significant meaning for other cell analysis studies beyond simply reviewing the cells classified as suspicious in prescreening.

Table 3 Confusion matix of prescreening test

Decision / Sample	Normal	Suspicious	Reject	Total
Normal	149	51 (25.5%)	0	200
Malignant	4 (2.8%)	195	0	200

(February, 1978)

6. ACKNOWLEDGEMENT

This development was sponsored by a grant from the Medical Information System Development Center in Japan.

7. REFERENCES

Holmquist, J., et al., "Computer Analysis of Cervical Cells. Automatic Feature Extraction and Classification," J. Histochem. Cytochem. 26:1000 (1978).

Moore, D. H., "Optimization of Cancer Detection," J. Histochem. Cytochem. 22:663-667 (1974).

Watanabe, S., et al., "A Pap Smear Prescreening System: CYBEST," in Digital Processing of Biomedical Images, New York, Plenum Publishing Corp. (1976), pp. 227-241.

Watanabe, S., and the CYBEST Group, "An Automated Apparatus for Cancer Prescreening: CYBEST," Computer Graph. Image Proc. 3:350-358 (1974).

APPENDIX

PARTICIPANTS

Dr. Kensuke Baba
Department of Pathology
School of Medicine
Dokkyo University
Mibu, Tochigi, JAPAN

Dr. James E. Green
Abbott Diagnostic Division
Abbott Laboratories
4757 Irving Blvd.
Dallas, Texas 75247, U.S.A.

Dr. S. Eiho
Automation Research Laboratory
Kyoto University
Gokasho, Uji, 611 JAPAN

Dr. H. Ishihara
Medical Systems Operations
Shimadzu Seisakusho Ltd.
Kyoto, JAPAN

Dr. Barry K. Gilbert
Biodynamics Research Unit
Mayo Clinic
Rochester, Minnesota 55901
 U.S.A.

Dr. Toru Kameya
Division of Pathology
National Cancer Center Research
 Institute
Tsukiji, Tokyo, JAPAN

Dr. M. Donald Graham
Coulter Electronics Inc.
Coulter Drive
Concord, Massachusetts 01742
 U.S.A.

Dr. T. Kaminuma
Tokyo Metropolitan Institute
 of Medical Science
Honkomagome Bunkyo-ku, Tokyo,
 113 JAPAN

*Non-participating co-author

Dr. J. Kariya
Tokyo Metropolitan Institute
    of Medical Science
Honkomagome Bunkyo-ku, Tokyo
    113 JAPAN

Dr. A. Kawahara
Research Laboratory
Nippon Kogaku K. K.
Nishioi, Tokyo, JAPAN

Dr. Kazumoto Kimura
Laboratory of Medical Sciences
School of Medicine
Dokkyo University
Mibu, Tochigi, JAPAN

Dr. H. Kitagawa
Automation Research Laboratory
Kyoto University
Gokasho, Uji, 611 JAPAN

Dr. Loren M. Krueger*
Biodynamics Research Unit
Mayo Clinic
Rochester, Minnesota 55901
    U.S.A.

Dr. S. Kurashina
Tokyo Metropolitan Institute
    of Medical Science
Honkomagome Bunkyo-ku, Tokyo
    113 JAPAN

Dr. M. Kuwahara
Automation Research Laboratory
Kyoto University
Gokasho, Uji, 611 JAPAN

Dr. Y. Masuda
National Institute of
    Radiological Sciences
Chiba University Hospital
Chiba, JAPAN

Dr. N. Miki
Automation Research Laboratory
Kyoto University
Gokasho, Uji, 611 JAPAN

Dr. K. Minato
Automation Research Laboratory
Kyoto University
Gokasho, Uji, 611 JAPAN

Dr. Kiyoshi Miyamoto
Laboratory of Medical Sciences
School of Medicine
Dokkyo University
Mibu, Tochigi, JAPAN

Mr. Philip E. Norgren
Optical Group
Perkin-Elmer Corp.
Norwalk, Connecticut 06856
    U.S.A.

Dr. Y. Okamoto
Toshiba Research and Development
    Center
Toshiba-cho, Saiwai-Ku
Kawasaki, 210 JAPAN

Dr. Keith Paton
Division of Medical Computing
Clinical Research Center
Watford Road, Harrow, GREAT
    BRITAIN

Prof. Kendall Preston, Jr.
Department of Electrical
    Engineering
Carnegie-Mellon University
Pittsburgh, Pennsylvania 15213
    U.S.A.

Dr. Richard Robb*
Biodynamics Research Unit
Mayo Clinic
Rochester, Minnesota 55901
    U.S.A.

Dr. I. Sasao
Toshiba Research and Development
    Center
Toshiba-cho, Saiwai-ku
Kawasaki, 210 JAPAN

Dr. Jack Sklansky
School of Engineering
University of California at
  Irvine
Irvine, California 92717
  U.S.A.

Dr. Stanley R. Sternberg
Environmental Research
  Institute of Michigan
Ann Arbor, Michigan 48107
  U.S.A.

Dr. I. Susuki
Tokyo Metropolitan Institute
  of Medical Science
Honkomagome Bunkyo-ku, Tokyo
  113 JAPAN

Dr. R. Susuki
Central Research Laboratory
Hitachi Ltd.
Kokubunji, Tokyo, JAPAN

Dr. S. Tamura
Department of Information and
  Computer Sciences
Faculty of Engineering Sciences
Osaka University
Toyonaka, Osaka, 560 JAPAN

Dr. K. Tanaka
Department of Information and
  Computer Sciences
Faculty of Engineering Sciences
Osaka University
Toyonaka, Osaka, 560 JAPAN

Dr. Y. Tateno
National Institute of
  Radiological Sciences
Chiba University Hospital
Chiba, JAPAN

Dr. T. Tomaru
Toshiba Research and Development
  Center
Toshiba-cho, Saiwai-ku
Kawasaki, 210 JAPAN

Dr. Saburo Tsuji
Department of Control
  Engineering
Osaka University
Toyonaka, Osaka, 560 JAPAN

Dr. S. Tsunekawa
Toshiba Research and Development
  Center
Toshiba-cho, Saiwai-ku
Kawasaki, 210 JAPAN

Dr. Masahiko Yachida
Department of Control
  Engineering
Osaka University
Toyonaka, Osaka, 560 JAPAN

Mr. Shinji Yamamoto
Central Research Laboratory
Hitachi Ltd.
Kokubunji, Tokyo, JAPAN

Dr. G. Uchiyama
National Institute of
  Radiological Sciences
Chiba University Hospital
Chiba, JAPAN

Dr. Yoichiro Umegaki
Division of Clinical Research
National Radiological
  Institute
Chiba, 280 JAPAN

Dr. H. Wani
Corporate Research and
  Development Operations
Shimadzu Seisakusho Ltd.
Kyoto, JAPAN

Dr. Sadakazu Watanabe
Toshiba Research and Development
  Center
Toshiba-cho, Saiwai-ku
Kawasaki, 210 JAPAN

AUTHOR INDEX

SUBJECT INDEX